Eric Priest was appointed a lecturer (1968) and professor (1983) at St Andrews University, where he founded an internationally renowned research group on solar theory. His research involves modelling the subtle interaction between the plasma atmosphere of the Sun and its magnetic field, responsible for much dynamic behaviour in the universe.

He sits on the Doctrine Committee of the Scottish Episcopal Church and is on the Board of Trustees of the John Templeton Foundation. Since 2007 he has helped organize a series of popular James Gregory public lectures on science and religion in St Andrews.

Honours include being elected a Member of the Norwegian Academy of Sciences and Letters (1994) and a Fellow of the Royal Society (2002). In 2002 he was awarded the Hale Prize of the American Astronomical Society and in 2009 the Gold Medal of the Royal Astronomical Society. In 2015 St Andrews University awarded him an honorary DSc.

He has edited 15 books, and written three research monographs and 460 journal papers. Hobbies include singing, playing bridge, climbing hills, keeping fit and enjoying time with his family (most recently twin granddaughters).

T0260760

REASON AND WONDER

Why science and faith need each other

EDITED BY ERIC PRIEST

First published in Great Britain in 2016

Society for Promoting Christian Knowledge
36 Causton Street
London SW1P 4ST
www.spck.org.uk

British Library Cataloguing-in-Publication Data
A catalogue record for this book is available from the British Library

ISBN 978–0–281–07524–9
eBook ISBN 978–0–281–07525–6

Typeset by Graphicraft Limited, Hong Kong
First printed in Great Britain by Ashford Colour Press
Subsequently digitally printed in Great Britain

eBook by Graphicraft Limited, Hong Kong

Produced on paper from sustainable forests

To my wife, Clare,
and children, Andrew, Matthew, David and Naomi,
with love

Contents

Contents

Contents

Contents

Illustrations

Contributors

Mark Harris is Senior Lecturer in Science and Religion at the University of Edinburgh. His first academic career was in physics, but after studying theology as preparation for ordained ministry, he became enthralled with biblical studies. After several years in chaplaincy and cathedral ministry he now combines his interests by running Edinburgh's master's programme in science and religion. He is interested in the ways that modern science has affected biblical interpretation, especially in understandings of creation and of miracle. He is the author of *The Nature of Creation: Examining the Bible and Science* (Acumen/Routledge, 2013).

Kenneth R. Miller is Professor of Biology at Brown University in Providence, Rhode Island. A cell biologist, he is the co-author of numerous biology textbooks widely used in US schools. He is also the author of the popular book *Finding Darwin's God: A Scientist's Search for Common Ground between God and Evolution* (HarperCollins, 1999). In 2014 he was presented with the Laetare Medal by Notre Dame University, an award described as the 'the oldest and most prestigious honour given to American Catholics'.

Michael J. Murray oversees the programme departments of the John Templeton Foundation. Before joining the Foundation, he was the Arthur and Katherine Shadek Humanities Professor of Philosophy at Franklin and Marshall College. Dr Murray received his MA and PhD from the University of Notre Dame. He is the author of *Philosophy of Religion: The Big Questions* (with Eleonore Stump) (Blackwell, 1999), *Reason for the Hope Within* (Eerdmans, 1998), *An Introduction to the Philosophy of Religion* (with Michael Rea) (Cambridge University Press, 2008), *Nature Red in Tooth and Claw: Theism and the Problem of Animal Suffering* (Oxford University Press, 2011), *The Believing Primate: Scientific, Philosophical, and Theological Reflections on the Origin of Religion* (with Jeffrey Schloss) (Oxford University Press, 2011), *Divine Evil? The Character of the God of the Hebrew Bible* (with Michael Rea and Michael Bergmann) (Oxford University Press,

2013) and *Dissertation on Predestination and Election* (Yale University Press, 2016).

David G. Myers is Professor of Psychology at Hope College in Holland, Michigan. His scientific writings, supported by National Science Foundation grants and fellowships, have appeared in three dozen academic periodicals. He has also digested psychological research for the public through articles in four dozen magazines and 17 books, including general interest books and textbooks for introductory and social psychology.

Eric Priest is Emeritus Professor of Mathematics at St Andrews University. A Fellow of the Royal Society, his research concerns the plasma physics of the dynamical behaviour of the Sun's atmosphere. This includes mechanisms for heating the Sun's corona to 1 million degrees centigrade and the nature of solar flares and huge eruptions that may interact with the Earth's environment. He has helped organize a series of James Gregory public lectures on science and religion in St Andrews since 2007. He is on the Advisory Board of the Faraday Institute and the Board of Trustees of the John Templeton Foundation.

Pauline Rudd is Research Professor of Glycobiology at University College, Dublin, and at the National Institute for BioProcessing Research and Training (NIBRT) in Ireland. She is also an Associate of the Anglican Community of St Mary the Virgin in Wantage, Oxfordshire. Her GlycoScience team sets out to define pathways involved in disease processes and to ensure the safety and efficacy of biotherapeutic drugs for cancer and autoimmune disorders. She has developed strategies for analysing glycans and discovering clinical markers for the diagnosis and treatment of cancers. Professor Rudd is a Fellow of the Royal Society of Medicine. She received a Waters Centre of Innovation Award in 2012 and an Honorary Doctorate at the University of Gothenburg in 2014. Before moving her group to Dublin in 2006, she was Reader in Glycobiology at Oxford University.

Jeff Schloss is Senior Scholar at BioLogos where he provides writing, speaking and scholarly research on topics that are central to the values and mission of BioLogos, and represents BioLogos in dialogues with other Christian organizations. He holds a joint appointment at BioLogos and at Westmont College, in Santa Barbara, California, where he holds the T. B. Walker Chair of Natural and Behavioral Sciences at Westmont College and directs Westmont's Center for Faith, Ethics and the Life Sciences. Schloss has a PhD in ecology and evolutionary biology from Washington University in St Louis, Missouri. He edited with Michael J. Murray *The Believing Primate: Scientific, Philosophical and Theological Reflections on the Origin of Religion* (Oxford University Press, 2011).

Eleonore Stump is the Robert J. Henle Professor of Philosophy at St Louis University, where she has taught since 1992. She has published extensively in philosophy of religion, contemporary metaphysics and medieval philosophy. Her books include her major study *Aquinas* (Routledge, 2003) and her extensive treatment of the problem of evil, *Wandering in Darkness: Narrative and the Problem of Suffering* (Oxford University Press, 2010). She has given the Gifford Lectures (Aberdeen, 2003), the Wilde Lectures (Oxford, 2006) and the Stewart Lectures (Princeton, 2009). She is past president of the Society of Christian Philosophers, the American Catholic Philosophical Association, and the American Philosophical Association, Central Division; and she is a member of the American Academy of Arts and Sciences.

John Swinton is Professor in Practical Theology and Pastoral Care in the School of Divinity, Religious Studies and Philosophy at the University of Aberdeen. He has a background in nursing and healthcare chaplaincy and has researched and published extensively within the areas of practical theology, mental health, spirituality and human well-being, and the theology of disability. He is the Director of Aberdeen University's Centre for Spirituality, Health and Disability: <www.abdn.ac.uk/sdhp/centre-for-spirituality-health-and-disability-182.php>. His publications include: *Spirituality in Mental Health Care: Rediscovering a 'Forgotten' Dimension* (Jessica Kingsley, 2001), *Raging With Compassion: Pastoral Responses to the Problem of Evil* (Eerdmans, 2007), *Living Well and Dying Faithfully: Christian Practices for End-of-Life Care* (edited with Richard Payne) (Eerdmans, 2009), *Disability in the Christian Tradition: A Reader* (edited with B. R. Brock) (Eerdmans, 2012) and *Dementia: Living in the Memories of God* (Eerdmans/SCM, 2012).

Keith Ward is a Professorial Research Fellow at Heythrop College, London, and an ordained priest of the Church of England. A Fellow of the British Academy, he was formerly Regius Professor of Divinity at Oxford University. He has taught philosophy and theology at a number of UK and US universities. His main work is a five-volume comparative theology, but he has also written books on science and religion from a philosophical perspective, most notably *God, Chance and Necessity* (Oneworld, 1996), *Pascal's Fire* (Oneworld, 2006) and *The Big Questions in Science and Religion* (Templeton, 2008).

David Wilkinson is Principal of St John's College, a Methodist minister and also Professor in the Department of Theology and Religion at the University of Durham. His background is research in theoretical astrophysics, where he gained a PhD in the study of star formation, the chemical evolution of galaxies and terrestrial mass extinctions. He also holds a PhD in systematic theology. His books include *God, Time and Stephen Hawking* (Monarch, 2001), *The Message of Creation* (IVP, 2002), *Christian Eschatology and the Physical Universe* (T&T Clark, 2010), *Science, Religion, and the Search for Extraterrestrial Intelligence* (Oxford University Press, 2013) and *When I Pray What Does God Do?* (Monarch, 2015).

Jennifer Wiseman is an astronomer and author. She studies the process of star and planet formation in our galaxy using radio, optical and infrared telescopes. She is also interested in science policy and public science engagement, and directs the programme of Dialogue on Science, Ethics, and Religion for the American Association for the Advancement of Science. She received her BS in physics from MIT, discovering comet Wiseman-Skiff in 1987, and continued her studies at Harvard, earning a PhD in astronomy in 1995. Dr Wiseman is a Fellow of the American Scientific Affiliation, a network of Christians in Science. She frequently gives public talks on the excitement of scientific discovery. She grew up on an Arkansas farm, enjoying late-night stargazing walks with her parents and pets.

N. T. (Tom) Wright read classics and then theology at Oxford. From 1978 to 1981 he was Fellow and Chaplain at Downing College, Cambridge. After a spell in Montreal, he returned in 1986 to Oxford as University Lecturer and Chaplain of Worcester College, Oxford.

He became Dean of Lichfield in 1994, Canon Theologian of Westminster Abbey in 2000 and Bishop of Durham in 2003. In 2010 he was appointed Professor of New Testament and Early Christianity at St Andrews University. Tom has published over 80 books at both scholarly and popular levels, most recently *Paul and the Faithfulness of God* (SPCK, 2013) and *Simply Good News* (SPCK, 2015). He has also broadcast frequently on radio and TV.

John Wyatt is Emeritus Professor of Neonatal Paediatrics at University College London. He has pioneered the application of optical techniques to the assessment of neonatal brain injury and was Co-Chief Investigator for the first major clinical trial of hypothermia as a practical treatment for babies who suffered birth injury. He has a long-standing interest in contemporary ethical debates about advances in medical technology and the beginning of life, and has frequently engaged in public and media debates about controversial issues in medical ethics. He is currently leading a research project into the philosophical and theological implications of advances in robotics and artificial intelligence. He is a member of the Ethics Advisory Committee of the Royal College of Paediatrics and Health, and the Medical Study Group of the Christian Medical Fellowship, and a board member of the Kirby Laing Institute for Christian Ethics.

Preface

Since December 2007 I have helped organize, in St Andrews, three highly popular James Gregory public lectures per year on science and religion, covering a wide range of topics in a high-level informative way that has appealed to many members of the general public. These have been generously funded by the John Templeton Foundation, and since November 2013 the main organizer has been Andrew Torrance.

Last year I was asked by the doctrine committee of the Scottish Episcopal Church to edit a booklet on science and religion and so invited nine of our previous and future James Gregory lecturers to contribute a short chapter. This was published as Grosvenor Essay number 11 in May 2015.

The present book is a greatly expanded version of the Grosvenor Essay and it includes three extra contributors. Each of the chapters stands alone as a real jewel of interesting insight, and I am very grateful to these eminent authors for giving their time and thought to contribute.

The chapters show how broad the field of science and religion has become and how deep the thinking is on the various aspects, from philosophy to astronomy, biology, human studies and theology. But the reader will notice many fascinating common themes and threads. Also, although the focus is mainly on Christian aspects, most of the points being made are common to the other Abrahamic faiths, which are after all closely related sister religions that worship the same God, but in different ways and with different insights.

All of the contributors share the belief that science and faith are not in conflict or independent but have complementary things to say, both being key to our human nature. But some of the chapters go further and suggest that they are not separate but share so many common features that they are better viewed as part of a common whole. In this more integrated approach, the notion of science as a monolithic concept is shattered after viewing the history of its development and asking 'What is it like to be a scientist in practice?'

Instead, the sciences and humanities represent a rainbow tapestry, merging into one another and linked by a common search for understanding, using reason and imagination.

Eric Priest
St Andrews

Acknowledgements

I am most grateful to the John Templeton Foundation for funding the James Gregory public lectures on Science and Religion at St Andrews and to Andrew and Alan Torrance, who have helped organize them. I am also thankful to the Scottish Episcopal Church for allowing us to create the current book by building on and adding to their Grosvenor Essay entitled 'Towards an Integration of Science and Christianity'.

1

Introduction: Towards an integration of science and religion?

ERIC PRIEST

The aim of this book is to bring together a series of world experts, who show in their own disciplines how many different aspects of both science and faith involve reason and wonder, which support one another and lead towards a much more integrated attitude to the sciences and humanities than is usually realized.

I am an applied mathematician (or theoretical physicist) and also a Christian, and in both I have been on a journey of discovery, or a pilgrimage, where my ideas have continually evolved. Not being an expert in theology or philosophy, this first chapter just represents some personal thoughts. In particular, I would like to challenge two views: the first is that the sciences are coldly inhuman and purely logical, whereas the humanities involve only our emotions and imagination; the second is that science is monolithic and reductionistic, governed by a simple set of instructions called the 'scientific method'. It is important also to counter the overspecialization that inhibits the natural human yearning to integrate and make sense of diverse knowledge.

This book takes some steps towards an integration of science and faith by moving away from a paradigm in which they are regarded as separate and having nothing to say to each other. As an introductory chapter, we build the case for developing an integrated approach by commenting on the possible relationships between science and religion, including the claim by New Atheism that they are at war (p. 2). There follows a discussion of the rise and fall of atheism (pp. 5–9) and a brief account of the way the words 'science' and 'religion' have evolved in meaning over the centuries (pp. 9–14). Then a personal insight into what it is like to be a scientist in practice (pp. 14–17) leads to a development of the argument for integrating the sciences and humanities, including religion (pp. 17–24), and for why science and faith need each other (pp. 25–30). Finally, a summary of the wide-ranging themes of this book is presented, from philosophy, through astronomy to evolution, biology, psychology and theology (pp. 30–9).

The relationship between science and religion

Ian Barbour (1997) suggested four possible relationships between science and religion, namely that they are:

1 in conflict
2 independent
3 in dialogue
4 integrated.

Conflict

The first possibility, that science and faith are at war or *in conflict*, is the one that has been stoked by the New Atheists. However, I have never personally felt a conflict between science and religion and suggest that such a conflict arises only if you misunderstand the nature of either science or religion. Thus, at one extreme you may have an *ultra-fundamentalist* view of religion with a wooden literalist interpretation of Scripture, which says you can 'learn your science from the Bible'. But this ignores the history of Christian ideas, in which St Augustine in AD 400 famously wrote: 'You should not inter-pret Scripture in a way that conflicts with reason and experience'. At the other extreme, *scientism* suggests that 'science produces the only reliable knowledge', but that is clearly false, since the questions that are most important to us as human beings are usually outside science, such as 'Am I in love?' 'Is that work of art beautiful?' 'What is my purpose in life?'

A clear philosophical argument against scientism has been pre-sented by Trigg (2015), who shows why science needs metaphysics. According to him, science does not have all the answers. It cannot explain why mathematics, a product of human minds, can unlock the secrets of the physical universe, nor why it can deal with abstract reasoning beyond the physical world. Indeed, scientists at the frontiers of physics happily contemplate universes beyond human reach. Thus, the foundations of science lie beyond science, and reasoning beyond the observable is needed to discover what is not yet known, so that metaphysics helps us conceive of realities apparently beyond our grasp.

Independent or in dialogue

In ancient Greece, the Stoics thought that God is everywhere, whereas the Epicureans believed that, if the gods existed, they took no interest in the world, but occasionally interfered. The latter led on to Stephen Jay Gould's (1989) idea of *non-overlapping magisteria* (see Fig. 1.1), which is similar to Barbour's relationship (2), in which there is no connection between science (the material world) and matters of religion (including ethics and morals) and so there is no possibility of conflict. However, this line of thinking naturally leads to a *deist god*, who is remote and uncaring and who is completely different from the Christian God. The next step, as science increases and you see no relevance for religion, is naturally to do away with religion altogether.

But the Christian God is very different from a remote deist god. The Christian God created the universe and all its laws with the potential to create life, and humans with a longing to study and understand the nature of God's handiwork. This God, however, is also intimately involved with the world, continually sustaining and supporting it. Thus, he or she is the continuous source of *all* our creative acts, whether they be in the sciences or the arts or our daily interactions with other people that are central to what it means to be human. Also, what Jesus shows is that God is *alongside* us in our suffering. This does not explain suffering, but it does mean that we are not alone, and is why I personally am much more attracted to Christianity than other religions.

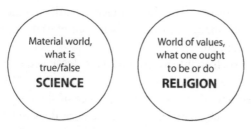

Figure 1.1 A diagram suggesting that science and religion are independent

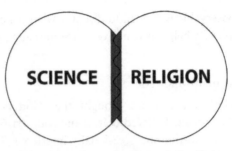

Figure 1.2 An alternative view of science and religion in dialogue

The third view, which I used to favour, is that science and religion intersect along a fuzzy boundary, where different questions are asked about the same reality and where they are in a respectful and listening dialogue rather than a state of war (see Fig. 1.2). The idea here is to recognize that some questions are scientific (often the *how* questions) and some are non-scientific (often the *why* questions) and that answers to both can be valuable (see pp. 14–15).

Integrated

A final model for the relationship between science and religion, and which we are moving towards in this book, is that they are integrated. There are several aspects to such a view, which are developed in this chapter. First of all, the history of science shows that the division into science and religion came about only in the late nineteenth century (see pp. 9–14). Second, science is not monolithic with a single approach but consists of a range of different fields that continuously merge into one another (see Fig. 1.3). Third, the idea that all scientists adopt the so-called 'scientific method' is a myth, since there are many approaches being employed in the different sciences (see pp. 20–2).

Aspects of such an integrated model are developed on pages 17–24 and 25–30. It has common features: with McLeish (2014), who proposes a 'theology of science' and a 'science of theology'; with Trigg (2015), who argues persuasively that science does not have all the answers but needs metaphysics; and with Bancewicz (2015), who shows how science enhances faith.

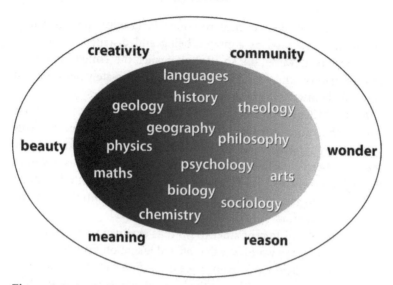

Figure 1.3 An integrated view of the sciences and humanities as a mixture of subjects (greys) that merge continuously into one another*

The rise and fall of atheism

Modern atheism is a complex mixture with several strands (e.g. Plant-inga et al., 2010). As science has developed and explained more natural phenomena, so a role for a *God of the gaps* has declined. So-called *scientific atheism* has grown, in which God is not needed for scientific explanation. In contrast, *humanistic atheism* was stimulated by the Enlightenment with its primacy of human reason and independence. Such atheism was articulated by Feuerbach (1804–72), with his view that our ideas of God are projections of our own minds, and reached a climax with Nietzsche's (1844–1900) proclamation of the *death of God*.

Modernism has, moreover, spawned at least two more types of atheism. One is *apathetic atheism*, with an indifference to the great questions of life, and a second is *protest atheism*, which is a cry against a God who seems indifferent to human suffering (e.g. Dostoevsky (1821–81) and Camus (1913–60)).

---

* The sciences and humanities have common aspects (creativity, beauty, wonder, reason and community) and are immersed in an underlying reality with meaning and purpose. In a more comprehensive, coloured version of the figure, the greys would be replaced by the colours of the rainbow and other disciplines would be added.

Atheisms are often critical reactions to classical theism, which emphasizes one God as supreme being and distant creator. Such reactions are, however, not so effective against the Trinitarian theism of Christianity, where God is a unity who is ultimately personal and continually interacting with his universe.

Alister McGrath (2004) describes the rise and fall of atheism. He stresses that atheism seemed for the last 200 years to be on the verge of eliminating religion as an outmoded superstition, but recently there has been a decline of disbelief and a rise of religious devotion throughout the world. He examines what went wrong with the atheist dream and suggests that religion is likely to play a central role in the twenty-first century.

He traces the history of atheism from its emergence in eighteenth-century Europe as a worldview that offered liberation from the rigidity of traditional religion to its golden age in the first half of the twentieth century. He exposes the flaws at the heart of atheism and argues that a renewal of faith is natural and inevitable.

The eighteenth century brought new ideas such as those of David Hume (1711–76) in the Enlightenment. A highly creative period of atheist reflection and critique of Christian ideas led to some seeing the Church as an enemy of progress. *Modernity* came into being by 1750 with its theme that reason is all that humanity needs to master the world and with atheism as its established religion.

In Catholic France, the French Revolution of 1789 aimed to over-throw tyranny and establish a rule of justice and equality, which signalled the dawn of a golden age for atheism. A temporary atheistic republic was set up with the only gods being the ideals of reason, liberty and fraternity. Voltaire was a deist, who felt that Christianity had corrupted a pure rational concept of God that could be known to everyone through reason or nature. Descartes (1596–1650) un-wittingly laid the ground for atheism by regarding true knowledge as being demonstrable and abandoning an appeal to experience. He set out to demonstrate the existence of God by philosophical argu-ment, proposing that God is the best explanation for the universe, but this led to D'Holbach suggesting that the universe can explain itself with no need for God.

The French Revolution encouraged the creation of a mindset of modern atheism by Ludwig Feuerbach (1804–72), Karl Marx (1818–83) and Sigmund Freud (1856–1939) as the presumption of the existence of God was replaced by intellectual scepticism. They argued in turn that religion is a human construction, that it dulls the pain of an unjust world, and that it is an example of obsessional neurosis, namely, the veneration of a father figure.

During the nineteenth and early twentieth century the perception grew of a conflict between science and religion (see pp. 9–14) in which religious belief would be eliminated as a relic of a superstitious age. However, the idea of an endemic conflict is no longer taken seriously by any major historian of science, and it is clear that there is no necessary connection between atheism and science, since some scientists are religious and some are not.

Gradually, a demand to prove beliefs grew, but it became clear to many that neither atheism nor Christian belief can be proved, and so Thomas Huxley coined the term 'agnostic' in 1869.

The event which converted much of the Western world to atheism was the publication of Charles Darwin's theory of evolution. One of the main reasons was that, within British Protestantism, a conception of God had grown up that proved later to be demonstrably in conflict with Darwin, namely, a way of defending God's existence. This defence had been proposed by William Paley (1802) and was widely accepted, including by the young Darwin. Paley suggested that many aspects of the world, such as the intricacy of the human eye and the human body, offered evidence of intelligent design; that is, of being planned in their present form by God, a benevolent watchmaker. Darwin's observations on board the *Beagle* of diverse populations of creatures and plants in different environments, as well as the fossil record suggesting that some species had died out, cast doubt on Paley's theory.

What is now clear is that Darwinism has no bearing on the existence of God. Some Darwinians (such as Asa Gray) are theists; some are not. Darwin himself changed from Christian to agnostic. The belief that scientists are necessarily atheists is a myth, as is the belief that, as a scientific worldview is increasingly accepted, the number of believing scientists will dwindle. A survey of the religious views of scientists in 1916 showed that 40 per cent have some form of personal religious belief, but, when the poll was repeated in 1996, the percentages were the same, with 40 per cent having active religious belief, 40 per cent having none and 20 per cent being agnostic.

The 'death of God', the loss of a sense of the presence of God in the face of so much injustice, culminated in the 1960s. Nietzsche suggested that God had ceased to be a presence in Western culture, so that there are no absolute values, and that belief had become a matter of taste rather than reason. The 1960s represented a crisis for Western Christianity, with the recognition that it had to renew itself or die; John Robinson's *Honest to God* (1963) dispensed with a God 'out there', while Bonhoeffer's letters from prison talked of *religionless Christianity*.

By 1970, many felt that religion was on the way out, secularism was triumphant and a more peaceful world was inevitable. But none of this proved true. Since then, atheism has lost its appeal. In Eastern Europe there has been a collapse of atheism and a resurgence of religious belief. In Western culture too, according to McGrath, we are in a post-atheist realm with atheism losing its cutting edge for several reasons:

1 Philosophical argument about God has declined with the realization that both Christianity and atheism are beyond rational proof. The idea that atheism is the only option for a thinking person is no longer widely held.
2 Whereas atheists used to argue that evil such as occurred at Auschwitz disproves the existence of God, many Christians would now reply that it demonstrates instead the depths to which humanity can sink and that God is a suffering God who is alongside sufferers.
3 Atheism produced a world without God that has been sucked dry of the imagination and creativity which are central to human nature and which urge us to press beyond the tangible to the spiritual; such longing and yearning for the transcendent is a natural part of being created by God.
4 The expectation of atheism was that faith in God would die out, but instead by 1980 a renewed interest in spirituality was evident.
5 A remarkable growth of Pentecostalism and a renewal of the mainstream denominations by the Charismatic movement produced a new sense of the presence of God.
6 Modernism had been replaced by postmodernism, characterized by a rejection of both absolute truth and intolerant atheism, and a tolerance for a variety of cultures and beliefs.

Whereas atheism at first appeared as a liberator from old ideas with a positive view of reality, it is now in a twilight zone, tarnished by memories of periods when it ruled as an intolerant oppressor.

Christianity is now being renewed and is growing across the world. In 1900 only 1 per cent of Koreans were Christian but now 49 per cent profess Christianity, having been attracted by its role as a liberator. In the West there is a growing awareness of the importance of spirituality in healthcare (see Chapter 11), and the role of religion in creating and sustaining a sense of community is recognized. Whereas the rise of atheism in the West was a protest against a corrupt and complacent Church, it has instead energized Christianity, encouraging it to reform itself.

History of the meanings of 'science' and 'religion'

In his excellent book, *The Territories of Science and Religion*, the eminent historian Peter Harrison (2015) asks, 'Have science and religion always been in conflict?' and gives a resounding 'no'. He suggests, for example, that claiming there was a war between Israel and Egypt in 1600 is clearly false, because neither state existed at that time. Similarly, claiming there has been a long-term conflict between science and religion is equally false, since the modern notion of 'religion' emerged only in the seventeenth century and that of 'science' in the second half of the nineteenth century. The idea that Christianity is a religion and that religions consist of beliefs and practices is a modern idea. Harrison's argument is as follows.

Scientia and *religio* began as *virtues*, inner qualities of an individual, personal habits of mind, one being an intellectual habit and the other a moral habit. Their meanings were later transformed into something quite different, namely, entities that are understood in terms of doctrines and practices. It is true that from the sixth century BC there was an attempt to describe the world systematically and to understand the principles behind natural phenomena, providing an account of causes in the cosmos, but this activity bears only a remote resemblance to the modern notion of 'science'. It is also true that societies have long engaged in worship in sacred spaces and had beliefs about gods, but it is only recently that the current notion of 'religion' set apart from secular domains has emerged.

These distinct notions arose because of political power and accidents of history, and are not natural ways to divide culture. It is a myth to suggest that science originated in Greek antiquity, declined in the Middle Ages, and finally broke away from religion in the seventeenth century and overcame religious prejudices, or that the Church hindered or ignored science. Rather, a religious perspective pervaded every area of Greek life, and ancient Greek accounts of the cosmos do not share the goals and methods of modern science.

Early concepts of *scientia* and *religio* (Aristotle, Augustine and Aquinas)

Greek philosophy entailed the pursuit of wisdom or happiness. For Aristotle (384–322 BC), all people by nature desire to know, and are assisted in this by acquiring intellectual virtues. For Plato all things in the heavenly order move according to reason, and a mathematical study of the heavens is a divine art that raises the mind of the philosopher: knowledge of nature enables such an individual to align his or her life with the rational principle that pervades the cosmos.

Aristotle distinguished three theoretical sciences:

1 *natural philosophy* (concerning what is finite, movable and inseparable from matter);
2 *mathematics* (what is immovable and embodied in matter);
3 *theology* (what is eternal, immovable and separate from matter).

Plato divided them instead into natural philosophy, logic and ethics. The heavens were studied because doing so promotes the development of moral and religious qualities or spiritual formation.

For Augustine (AD 354–430), *religio* is rightly directed worship. Believing in God meant loving him, and 'doctrine' meant teaching or interpreting Scripture rather than a systematic philosophy. The idea of 'believing in the existence of' began in the seventeenth century. Natural philosophy was always pursued with moral and religious ends in mind. There was no opposition between naturalistic and religious accounts of the cosmos, since the 'books' of nature and of Scripture are both modes of divine communication. As Augustine famously said, 'You should not interpret Scripture in a way that conflicts with reason and experience'. In the early Middle Ages, there was a belief in both a literal and a spiritual or allegorical sense of Scripture, and the intelligibility of nature lay mainly in its moral and theological meaning. Science was a mental habit rather than a body of nature.

For Thomas Aquinas (1225–74), *religio* refers to inner piety, interior acts of devotion and prayer, whereas *scientia* is a habit of mind, an intellectual virtue, a personal quality. Rehearsing well-known scientific knowledge strengthens the mind. He stressed three virtues:

1 *understanding* (grasping first principles);
2 *science* (a mental habit gradually acquired by deriving truths from first principles;
3 *wisdom* (grasping the highest causes).

Aquinas departed from Aristotle in including in *scientia* the seven gifts and nine fruits of the Spirit, such as faith, hope and love. Thus, *scientia* is a personal quality with a moral component. In the Middle Ages it included the study of grammar, logic, rhetoric, arithmetic, astronomy, music and geometry, and Aquinas suggested it should also include Aristotle's disciplines of natural philosophy, mathematics and theology. Natural philosophy included the study of God and the soul, but excluded mathematics and natural history.

Reformation and scientific revolution (sixteenth to eighteenth centuries)

From the sixteenth century, because of the Reformation and the rise of experimental natural philosophy, a new conception of *religio* arose as a set of beliefs and practices rather than an inner piety. Thus, belief now entailed an act of giving intellectual assent to propositions rather than trust in a person. Also, Protestants criticized the idea that virtues can be achieved through practice (because of original sin and the need for grace) and felt that individuals are justified not by an inner quality but by their relationship with God (as emphasized by Luther).

In the seventeenth century, the term 'science' began to refer to *methods and laws of nature* with regularities imposed directly by God. This replaced Aristotle's inherent tendencies and powers of natural objects. Whereas previously, scientific knowledge was an instrument for producing a scientific habit of mind, now a scientific habit of mind became an instrument for producing knowledge.

The earlier idea of *progress*, referring to personal progress towards virtue, was replaced by a new concept associated with usefulness and referring to a cumulative addition to an external body of knowledge and facts. For Bacon, Descartes and Boyle, religion and science restore mastery over oneself and over nature, respectively, so the purpose of natural philosophy becomes one of producing useful knowledge and having command of nature.

The idea of religion and science referring to *activities and bodies of knowledge* in turn led to a growth of *natural theology*, as arguments for the existence of God were sought based on reason and the ordinary experience of nature. Francis Bacon (1561–1626) proposed a new non-allegorical way of reading the 'book of nature' – a Christian approach to nature in which God shows his omnipotence and wisdom but not his image. Kepler (1571–1630) and Galileo (1564–1642) said this book is written in the language of mathematics, while Descartes (1596–1650) talked about natural laws imposed on matter by God and capable of mathematical formulation. For Newton (1642–1727) 'the system of planets could not have arisen without God' and 'the regularities of nature show the continuous action of God' (Newton, 1999 [1687]). Boyle sought the meanings of the book of nature instead by experimentation, dissection and magnification.

All this represented a new partnership between theology and the new science. A rich interdisciplinary culture existed, held together by natural theology, which was integral to the scientific endeavour, so that discourse about God was a genuine part of natural philosophy. A unity of knowledge was assumed, so that truths about the natural world could not in principle be in conflict with religious truth. Christianity motivated science by stating that nature is orderly, good and worthy of study.

However, a collapse in the distinction between natural and supernatural causes meant that nature has either divine causes or natural causes. Similarly, meaning was either symbolic or literal. Eventually, this flattening of the scope of causation and meaning meant that theology and science came to occupy the same explanatory territory, which led for some to a competition between them since they both became concerned with knowledge. At the same time, natural philosophy shifted its focus away from contemplating truth to providing human comforts and useful knowledge, and to mastering nature.

Professionalization of science (late nineteenth century)

Whereas science had in the seventeenth century begun a transformation towards its modern meaning, this was only completed in the late nineteenth century due to the professionalization of science, the redefinition of 'science', and the invention of the terms 'scientist' and the 'scientific method', and the myth of 'a perennial conflict between science and religion'. This coalescence of science meant that it was possible for the first time to speak of science and religion as two independent activities with a relation between them. It also led to a transfer of authority from religion to science. T. H. Huxley (1863), for example, claimed that the scientific method is the only way to ascertain truth.

The professionalization of science involved a determination to separate science from amateur activity and to put clear bounds around it. Thus, 'science' was now defined to be natural philosophy plus natural history, but excluding theology, metaphysics and the human-ities. By 1880, the term *natural sciences* was preferred to 'natural philosophy', and *biology* had replaced 'natural history', since it was regarded as a more scientific and serious occupation. This severing of ties with the humanities has split intellectual life unhealthily into *two cultures* (Snow, 1959).

'Science' was reconstructed around the principles of a common method (which excluded religious and moral considerations) and a common identity of its practitioners. The word 'scientist' was coined by William Whewell in 1847, which marked such a person out as distinctive. In the same year, the Royal Society, which had been previously dominated by gentleman amateurs, tightened its rules to ensure its fellows had genuine scientific accomplishments. As a result, the number of fellows had almost halved by 1899 and the participation of Anglican clergymen had fallen by two-thirds.

The invention of the so-called 'scientific method' was another factor that gave a coherence and unity to science. Bacon had pro-posed the method of induction, whereby observations are catalogued in the hope that a pattern will emerge. But by the end of the nine-teenth century the humble accumulation of data had been replaced by the work of the scientist armed with the scientific method, in which experiments were designed to adjudicate between different hypotheses.

The 'conflict myth', alleging that there has always been a conflict between science and religion, was invented in order to consolidate the boundaries between them in two books, by Draper (1874), who was a chemist and amateur historian, and White (1896), who was the first president of the first secular university (Cornell). These writers claimed that:

- medieval thinkers thought the Earth is flat;
- heliocentrism was resisted because it demotes humans from the centre;
- Darwin met with universal resistance from the Church;
- the Church banned human dissection;
- the Church opposed vaccination and anaesthesia;
- Galileo was tortured and imprisoned by the Inquisition.

These are all sheer fabrications.

To take the most celebrated example, while it is true that Galileo was tried by the Inquisition and forced in 1633 to recant the Copernican hypothesis (that the Sun lies at the centre of the solar system), this was basically a conflict not between science and religion, but within science, and between an arrogant Galileo and his former friend the Pope, whom he had ridiculed in a book. The Church was in fact the strongest supporter of astronomy and was simply accepting the standard scientific consensus of the period. The Church was happy to accept heliocentrism as a hypothesis, but Galileo insisted it is a fact, even though he couldn't prove it and the evidence was flimsy. Indeed, good support for the idea only appeared 50 years later with Newton's *Principia* (1687), and two important consequences were first demonstrated much later, namely, parallax in 1838 and the rotation of the Earth in 1851.

What is it like being a scientist?

In most areas of science, we cannot find absolute truth or prove that a theory or model is correct, but we can ask, 'Is it *consistent* with our observations?' The aim then is to improve the model as time goes on and so come closer to the truth. In a similar way, I can't prove that God exists, but I can ask, 'Is God's existence or non-existence more consistent with my experiences?' For me, as a Christian, the *existence* of God is more consistent, and so I have chosen to live my life for the time being with the assumption that God does exist: trying to follow the example of Jesus in my life; studying the Bible; being part of a Christian community; and listening to the promptings of the Holy Spirit.

Different types of question

For any question, it is important to identify whether it is a scientific or non-scientific question. Thus, 'Is the Earth warming?' and 'How did *Homo sapiens* arise?' are scientific, but 'What should we do about climate change?' and 'Does God exist?' are non-scientific questions.

Different questions can be asked about the same event. Thus, 'How or why is the kettle boiling?' may have one answer in terms of physics and another in terms of my wife's thirst. Again, 'How or why are two people kissing?' could have as one answer 'By the application of suction during the anatomical juxtaposition of two orbicular oris muscles in a state of contraction', but another answer is likely to be of much more interest to those who are romantically inclined.

For a scientific question, it is also important to determine whether it is part of *mainstream* science, in which case the answer is widely accepted and is unlikely to change, or is part of newer, more specula-tive science, which is on the fringes of knowledge and cannot yet be trusted. If a question is part of mainstream science, then we should (at least provisionally) accept and trust the answer given by the experts. Examples here include 'Did the universe arise from a Big Bang 13.8 billion years ago?' or 'Did humans arise by evolution with natural selection?' or 'Has the Earth's global temperature risen by 1 degree centigrade in the past 100 years?' Furthermore, in view of the huge range of different sciences (see pp. 19–20), authoritative answers to these questions cannot be given by any old scientist but only by the *experts* in that particular branch of science. Thus, just because someone is a 'scientist', it does not mean that that person's views on any topic outside his or her expertise can necessarily be regarded as of value. It is only climate scientists, for example, who can talk with authority about climate change.

Being a scientist

So, what is it like to be a scientist? Is scientific work cold, rational, logical, mechanical, undertaken by computers and emotionless people in white coats, and having nothing to do with the arts or Christianity? Is the world reductionistic, with the weather determined by individual clouds, life governed by individual molecules in cells, people determined by their genes, and our thoughts determined by individual electrical signals in neurons?

Modern science is far from being clockwork and deterministic (see Chapter 3). It involves a combination of regularity and chance, of law-like and random behaviour. Statistical fluctuations are common, and so laws can only predict in general terms, but are unable to determine well in advance the formation of, say, individual cyclones or sunspots or stars. The nature of time and space and matter at the most fundamental level so far explored is also full of uncertainty. In addition, science often deals in multiple levels of description, with the lower levels affecting the higher and the higher feeding back down to the lower, so that one cannot predict the higher level of behaviour by studying the lower level alone (see pp. 19–20). Examples include the behaviour of the weather, flocks of birds and a human body.

Being a scientist in practice from my experience involves:

1 *creativity*, leaps of faith, intuition and imagination, in which inspiration is followed by perspiration as the skills honed over many years are used to work out an idea;
2 a sense of *beauty* and *wonder* and therefore *humility*, and so, if scientists behave in an arrogant manner, they are not being true to their science;
3 openness and *questioning*, which lead to a voyage of discovery;
4 *trust* and integrity, which are crucial for the scientific *community*.

The crux here consists of the wonderful moments of inspiration, which Tom McLeish describes as follows:

> With a spark of insight, a new light shines on the previously dark part of the unknown world as something almost unspeakable happens. This is what we work for. Those miraculous moments when the fog clears and we know something for the first time really are 'more precious than rubies'. (McLeish, 2014, p. 177)

Being a scientist affects in a profound way my life of faith, and so being a Christian for me involves:

1 creative leaps of *faith*;
2 a sense of *beauty, wonder* and *humility*, which often arises;
3 openness and *questioning*, which lead to a pilgrimage;
4 *trust* in the *community*, the body of Christ.

The close parallel between the nature of being a scientist and a person of faith suggests to me an underlying unity. Thus, a scientist can indeed be a Christian, provided he or she is open to the insights of science and is responsive to the hand of the maker in the universe.

In his wonderful book *Faith and Wisdom in Science* (2014), Tom McLeish has captured the spirit of curiosity and wonder in the sciences by stressing that wisdom has a long Greek and Hebrew history, and that the modern word *scientist* (a person who 'knows') has a less appealing feel than the earlier *natural philosopher* (a 'lover of wisdom'). He regards the sciences as 'searching for the wisdom of natural things' (McLeish, 2014, p. 25) and points out that it is more about imagination and creative questioning than about method or logic, so that wisdom and the understanding of nature speak of a deeper significance than simply knowing things.

In her recent book *God in the Lab: How Science Enhances Faith*, Ruth Bancewicz (2015) brilliantly presents many examples from the stories of other scientists that echo my own experience described above (of creativity, imagination, beauty, wonder and awe as a scientist).

A more integrated view of science and religion

From the summary on pages 9–14 of Harrison's account of the historical changes in the meaning of science, it was clear that the sciences do not share common features that enable us to speak of one 'science'. Harrison (2015) suggests that there is instead a cluster of myths that sustain our idea of science, so that in reality it is better to speak of 'sciences'. He also suggests that science and religion are not necessary features of human society or natural ways of dividing human activity. Rather, they are ways that are peculiar to Western culture and are a consequence of particular historical circumstances. He notes that we often perpetuate conflict by conceding the cultural authority of science, as well as the propositional nature of religion and the notion of a neutral rational space where dialogue can take place. So here I offer some thoughts about how to develop an integrated view of the sciences and humanities, starting with a parable and a discussion of the different levels of the sciences. I show that the scientific method is a myth and then go on to offer a suggestion for a new unifying feature of the sciences and humanities, including religion.

A parable

Hyung Choi (private correspondence) suggested a clever parable concerning two islands labelled 'science' and 'religion' with their own languages and terminology and poking up above a thick mist (see Fig. 1.4a overleaf). Some people tried to build a rickety bridge between the two islands, but a deeper truth was revealed when the Sun came

Figure 1.4a Science and religion viewed as separate 'islands' obscured by mist

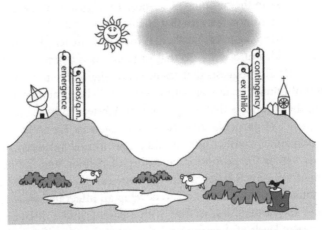

Figure 1.4b Science and religion viewed as 'one island with two hills' joined by dry land

from behind the clouds and the mists cleared to show that the two islands were not separate after all but were joined by dry land (see Fig. 1.4b).

Perhaps there is indeed one island but different maps, one underlying reality, one building but different drawings, one truth illuminated from different directions. But each map or drawing or direction is incomplete and needs the others for a fuller understanding.

Different levels of understanding

Science is not monolithic with a uniform method and understanding, although in common talk it often sounds as if it is. Rather, the many different sciences that form the (greyscale) rainbow of Figure 1.3 represent different levels of understanding, as illustrated schematically in Figure 1.5. At the lowest level, we have the subatomic world of particle physics, governed by quantum mechanics, which describes the inner structure of atoms – or, rather, this is the lowest level we know about for now, since there may well be other lower, more fundamental, levels that will be revealed in future. Above it lies kinetic theory, which describes the motions of atoms as they move around and interact with one another.

Gases, liquids and plasmas consist of individual particles, but for many purposes they can be treated as continuous media, described by the equations of fluid mechanics and magneto-fluid mechanics. These in turn can describe the behaviour of our Sun or other stars and of galaxies in the field of astronomy.

However, a separate line of development from atoms is to combine them together as simple molecules and compounds, described by

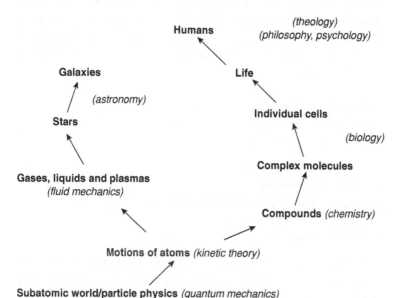

Figure 1.5 Different levels of understanding in the sciences (other sciences could be added for completeness)

chemistry. In turn, this knowledge is developed to describe complex molecules, and in biology the properties of individual cells and of primitive life. Complex life leads on to humans, where new fields (such as philosophy, psychology and theology) appear and are necessary in order to describe their distinct features.

Each of these different levels is described by its own science, with its own terminology and models of behaviour. The complexity is quite staggering. Thus, 1 cubic centimetre of the air in the environment where you are reading this book contains not 1 million (10^6) or even 1 billion (10^9) or 1 trillion (10^{12}) particles, but 10 billion trillion (10^{22}). Again, in a single galaxy there are 100 billion stars (10^{11}), so that in the whole visible universe there are 100 billion trillion (10^{23}). In the human brain there are 10 billion nerve cells (10^{10}), each with 5,000 connections on average, whereas in the human body there are in total 37 trillion cells.

Modern sciences, however, are not reductionistic, so that you often cannot deduce the higher-level properties from the lower-level ones (see Chapter 3). The higher-level properties *emerge* from interactions of the simpler lower-level structures, and these *emergent properties* can interact back down and affect the lower levels. Examples of emergent structures are snowflakes, or flocks of birds, or a mound of termites, or consciousness emerging from electrical signals in neurons, or human properties emerging from the interactions of many, many individual cells.

The scientific method

The so-called 'scientific method' is commonly described in terms of several steps (see Fig. 1.6):

1 asking a question;
2 stating a hypothesis;
3 conducting an experiment to test the hypothesis;
4 analysing the results;
5 making a conclusion.

Then the conclusion is used to ask another question or to refine the hypothesis or the experiment, and the process is repeated until the results converge towards a provisional or final conclusion.

The method was proposed by Peirce (1878), building on many previous ideas by Aristotle, Al-Haytham, Ibn Sina, Grosseteste, Roger

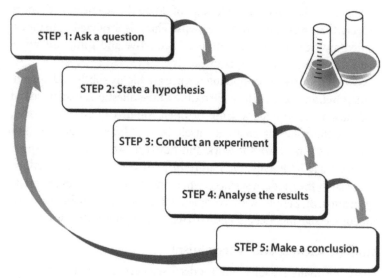

Figure 1.6 Steps in the so-called 'scientific method'

Bacon, Francis Bacon, Descartes, Galileo, Newton and Hume, and was later developed by Popper and Kuhn. However, the sciences are nowhere near as simple as this description would imply. As a single scientific method, the process is actually a myth, which is sometimes followed by scientists, but usually not.

Thus, let us consider each of the steps in Figure 1.6. The first is not necessary, since sometimes a scientist may, rather than start with a question, stumble on a discovery by accident or when he or she is pursuing a different line of reasoning. Also, in some cases the second step is irrelevant, since one may start not with a hypothesis but with an experiment or by gathering data and seeking a pattern in them. Furthermore, experiments are not always involved, since they may be impossible: in, for example, astronomy you cannot conduct a real experiment on, say, the formation of stars, and so instead you gain knowledge by making observations and developing theories. Again, some experiments may not be affordable (e.g. going to very much higher energies than is possible with the current Large Hadron Collider), while others may be unethical (e.g. in parts of medicine) or inappropriate (e.g. in parts of geology or psychology, where you gather data instead).

There are thus many diverse ways of doing the sciences, with more aspects than just the above five steps. There is a range of methods, some empirical and some not, some deductive and some inductive, just as in many of the humanities.

Several features of conducting science are highly subtle and deserve in-depth discussion, such as the nature of the questions that are asked and how to encourage a sense of open-minded questioning. Also, there is the topic of creativity itself in the different sciences and humanities – how does it work and how can we develop the space to encourage it? Finally, the nature of personal human interactions is crucial in the modern sciences, where most great advances are made by groups of people with complementary skills – so how do we encourage sharing, debate and co-creation in an effective way?

Modern sciences and determinism

There are many reasons why modern sciences are often not deterministic. One is the presence of *random statistical fluctuation* and *turbulence*, which involve a combination of regularity and chance or of law-like and random behaviour. For example, in the Sun's atmosphere or the Earth's weather, you can often predict the general properties, but not the appearance and behaviour, of individual sunspots or cyclones. The inherent uncertainty in the weather means that there is probably an upper limit in terms of a week or two, beyond which it will never be possible to predict the weather in places where the weather is highly variable such as the British Isles.

Randomness is common at many levels of understanding. Thus, for a fluid picture of the behaviour of a continuous medium, such as air or solar plasma, *turbulence* is common, where you need to employ statistical methods to determine the mean or probable behaviour. If you go down a level to a particle picture and model the many millions of air atoms or particles in the solar atmosphere, you can determine only the *probable distribution* of particles and their mean behaviour. Again, going down a further level to a quantum picture, there is no independent objective reality; position and velocity are *uncertain*, according to Heisenberg's uncertainty principle, so that it is only their product that is well determined.

Chaos is another common feature, where small changes in initial conditions can lead to large changes in the resulting behaviour. Poincaré first studied chaos in the motion of three particles under gravity in the 1880s, but the modern development started with Edward Lorenz's (1963) famous paper on a very simple model of weather consisting of three ordinary differential equations that exhibit chaos. A simple example of chaos is a double pendulum, where you attach one pendulum to another so that it hangs under gravity below the first. A single pendulum exhibits regular behaviour, such that if you start it going with a slightly different position or velocity, the resulting oscillation will just be slightly different. However, a double pendulum can exhibit chaos, so that a small change in the initial conditions can lead to huge differences in behaviour. An interesting aspect is that sometimes order can emerge from chaos.

Another aspect is that the nature of time and space and matter at a fundamental level is uncertain. Is space and time continuous or granular? What is the nature of time – is the flow of time continuous or discrete? Is there an objective time or does it depend on the observer? Is matter like a series of Russian dolls that go on getting forever smaller or is there a fundamental particle of which quarks are composed?

Integrating the sciences and humanities

On pages 2–5 we summarized the different models for relating science and religion, dismissing the idea of a conflict as being a superficial reaction that does not appreciate fully the nature of the sciences and religion. Whereas the second idea, that they are independent, may appeal to those who like to categorize subjects neatly into boxes, it does not do justice to the rich interplay between the two and the common aspects that many of those of us who practise science feel (see pp. 15–17). So a natural approach is the third, namely, that the sciences and religion are in dialogue.

However, in this book we are moving towards a more holistic idea, namely, that they are integrated. There are several aspects to such a view. First, the history of science reveals that a division into science and religion came about only in the late nineteenth century (see pp. 9–14), largely because of the professionalization of science. This defined 'science' as being natural history plus natural philosophy, in order to put bounds around it and separate it from amateur pursuits and the humanities. At the same time the word 'scientist' was invented, to describe a professional who used the so-called 'scientific method', but this is a myth, since scientists in practice utilize a wide range of different techniques (see pp. 2–22).

Tom McLeish (2014) proposes similar ideas to our own. He stresses that science and theology take the entirety of nature as their subject, since there really is one world, rather than two *magisteria*, so that our minds are the locus of both meaning and explanation within it. For him, science is not value-free and values are not science-free, since we need to know why we do science, what purpose lies behind our great questions, and where science lies in the stories that bind our culture and tell of our history. He proposes that, rather than speak of 'science and religion', we should build both a 'theology of science' and a 'science of theology'. Science and theology are not in combat or complementary; they are 'of each other'.

The integrated model that we have in mind is sketched in Figure 1.3. The core idea is that science is not monolithic with a single approach, but consists instead of a range of different sciences that continuously and seamlessly merge into one another and into the humanities (see pp. 19–20). Both the sciences and the humanities are immersed in an underlying unity that shares the same features of creativity, community, beauty and wonder (see pp. 15–17), and which provides meaning, where different questions are asked about the same reality. Attending inaugural lectures in the whole range of academic studies over the years, I have been struck time and again how inspiration and creativity are crucial to the sciences, as new professors have described how they were led along unexpected paths in response to hunches and to vibrant new ideas. But I have been equally astonished at how clever academics in the arts follow reason, logic and patterns of understanding and behaviour as well as inspiration.

In this more holistic approach to the sciences and humanities, what united them in classical times and before the nineteenth century was Aristotle's idea of virtues, such that *scientia* and *religio* are different kinds of virtue (see pp. 10–11). So what unites them today? In my view, it is a common *search for understanding*, which involves both *reason* and *imagination*. They use a whole range of different types of question and often lead to a sense of wonder and beauty. It is not the case that the sciences are purely a rational exercise whereas the humanities give free unconstrained rein to the imagination. Rather, both reason and imagination are central to the sciences and humanities, to both types of endeavour and exploration.

Why science and faith need each other

Reason and imagination

We have suggested that the sciences and humanities are part of an integrated whole that combines *reason* with *imagination*. If it is true that they are part of one family, then it is important for different members of the family to respect each other and to recognize the wide range of complementary gifts and insights that each may offer.

We have argued that, in general, reason and imagination need each other. Without *rationality* and experience, one loses touch with reality and is unconstrained, wandering around aimlessly. Thus, a theoretical scientist without reason and observation can dream up any explanation that bears no relation to reality. Theorists such as myself need to be grounded in observations or experiment. Although reason has long been recognized as crucial to scientific endeavour, it is also crucial for the arts, since it is the constraints of a rational framework that produce the pressure and the desire to be creative and lead to the construction of patterns that are recognized as essential in great art or music.

Without *creativity* in the sciences, on the other hand, the path of reason soon stops. Indeed, uncertainty and doubt are essential for creativity and can drive one to discover new and unexpected results. Imagination has always been recognized too as being central to the arts, as the source of creative acts of painting or music, for example, but it is the constrained working out of a creative idea that leads to a final result.

Sciences and faith

However, how do the sciences and faith influence one another? Why do they need each other? Our hearts and minds are not opposites at war with one another, but are complementary parts of what it is to be human. God gave us both and so we need to value both and to hold them together in a healthy balance. Thus, the sciences, the humanities and faith are part of an underlying unity, which we glimpse when we open our eyes in humility to the wonder and beauty of the sciences and to the hand of the maker in the universe.

For me, my science affects faith. Studying the Sun often gives me a sense of wonder and beauty for two reasons:

- the amazing structures we observe in the Sun's atmosphere;
- the underlying elegant mathematics for the behaviour of plasmas and magnetic fields – it is there waiting to be discovered.

The fact that mathematics underpins the universe is consistent with there being a divine creator. The symmetry and elegance in the mathematics is uncanny and points, for me, to a deeply rational heart to the universe that is reflecting an aspect of the divine mind.

Faith also affects the sciences. It underpins the sciences, gives a profound reason to study them and provides an important support for good science. If the universe is the result of God's willing it into being and if God has given us minds to appreciate and to long for understanding, it is a natural privilege for us to learn about the universe and to explore the way it works, from distant stars and galaxies to the inner workings of cells and the nature of space, time and matter. The historical roots of modern sciences lie in the conviction from faith that the exploration of the natural world is an act of praise and worship. Indeed, the sciences are based on two elements of faith, namely, that the universe can be understood by rational enquiry and that knowledge of it from science is preferable to ignorance.

Faith encourages positive attitudes and virtues in scientists, such as humility, integrity, wonder, respect, appreciation and, especially in modern scientific groups, teamwork and community. Indeed, religion stresses the importance of relationship, which is at the core of the Christian Godhead and is so important in the effective working of the contemporary sciences.

There is a continuing need for science to have an underlying narrative – a moral or aesthetic vision to promote its relevance. In the past, natural theology fulfilled these functions for natural history and natural philosophy. In the 1950s and 1960s, science was seen as the engine of progress and wealth, but this pure vision has been shattered by the negative global aspects of pollution, climate change and war. The sciences do not have the motivation within themselves to provide answers to such global questions. They have the potential tools, but it is only if faith and moral vision can provide the right motivation that the sciences can be mobilized effectively by society.

Thus, for example, we need to recognize that we are not completely independent individuals whose only motivation is our personal happiness or the profit motive. Rather, we are an integral part of God's creation with a special responsibility to care for it. We are also part of the global humanity, of a common world community that is bound together by our concern for each other.

In his book *The Great Partnership*, Jonathan Sacks has given the most persuasive argument I have come across for a partnership and integration between the sciences and faith in a search for *meaning*. The sciences speak with expertise about the future and invoke the power of *reason*, whereas religion speaks with authority about the past and invokes the higher power of *revelation*. Thus, it is the creative tension between the two that keeps us sane, such that the sciences are needed to ground us in physical reality, whereas religion preserves our spiritual sensibility, keeping us human and humane. A balance is needed, with the sciences 'taking things apart to see how they work' and religion 'putting things together to see what they mean'. In other words, there is an integration of left-brain and right-brain activity. As Einstein (1954, p. 46) said, 'Science without religion is lame, religion without science is blind.'

In an inspired section, Jonathan Sacks compares both contributions as follows:

> Science is about explanation. Religion is about meaning. Science ana-
> lyses, religion integrates. Science breaks things down to their component
> parts. Religion binds people together in relationships of trust. Science
> tells us what is. Religion tells us what ought to be. Science describes.
> Religion beckons, summons, calls. Science sees objects. Religion speaks
> to us as subjects. Science practises detachment. Religion is the art of
> attachment, self to self, soul to soul. Science sees the underlying order
> of the physical world. Religion hears the music beneath the noise.
> Science is the conquest of ignorance. Religion is the redemption of
> solitude. (Sacks, 2012, p. 6)

He goes on to stress that the *meaning* of a system lies outside itself, so the meaning of the universe lies outside the universe. Thus, when the transcendental God of monotheism was discovered, the God who stands outside the universe and creates it, it became possible to believe that life has a meaning. Indeed, it is meaning that creates hope, human freedom and dignity.

To the question 'Why are we here?' the answer from the sciences is silence, since the search for meaning has nothing to do with the sciences. The idea of sheer happenstance and that the purpose of humans is to blindly replicate themselves is a barren and bleak one. Rather, the discovery of God is the discovery of meaning, a God who creates us out of a selfless desire to make space for otherness – in a word, a God who loves. We exist because we are not alone – we are part of a cosmic drama of relationship.

Insights on science and faith from other chapters of this book

Keith Ward points out that scientific explanations refer to physical data, but not all explanations are scientific. In particular, physical sciences do not ask whether anything is of *intrinsic value*, and they set aside questions of consciousness or subjective feelings, being wary of speaking of purposes or goals. However, God created the world for the values it realizes that can be enjoyed by God and also by humans. Again, God gives a good explanation for the laws of nature, since they give rise to a universe with good and valuable purposes.

Astronomy strongly suggests elegance and order in the laws of nature, and beauty, complexity and variety in the universe, but science needs faith in order to answer whether or not there is *purpose* in the cosmos. Faith in God suggests there is an intrinsic value in beauty, intellectual understanding, creativity and personal relationships, and so encourages us to appreciate them.

David Wilkinson stresses that science cannot explain the role of God at the beginning or end of the universe. For him, science and religion have different foci in exploring the universe, but there are areas of overlap where each can enrich the other and indeed pose questions for the other. One area where faith can go beyond science is in addressing the question 'Where do the laws of physics come from?' The Christian doctrine of creation sees God as the fundamental basis for the origin and continued existence of the laws by which the universe evolves. Such a God is immanent and transcendent in sustaining all things. This insight gives us hope of a new creation in the face of the cold, desolate future that science predicts for the universe. It also motivates us to use the sciences for purposes of goodness, justice and love.

For Kenneth Miller, biological science implies that the capacity for evolution is built into the structure of the natural world as an inherent part of the physics and chemistry of matter. Thus, the appearance of humans is not a random accident but a direct consequence of the characteristics of the universe. Evolutionary science suggests that we are part of the grand, dynamic and ever-changing fabric of life that covers our planet. What faith adds is the concept that we live in a universe of meaning and purpose, and so an understanding of evolution deepens our appreciation of the scope and wisdom of the Creator's work. To the person of faith, evolution is God's method of creation.

Pauline Rudd argues against biological determinism by stressing that at the biochemical level we are much more than our genes. She also points out that the sciences are ethically neutral, and that faith can help us decide how to live. She concludes that life can be lived without in-depth knowledge of the sciences and religion, but how much richer it is to explore the life-enhancing complementary aspects of both.

David Myers points out the importance of holding untested beliefs tentatively, assessing others' beliefs sceptically and using the sciences to winnow truth from error. Such ever-reforming empiricism has changed his mind on many occasions. There are striking parallels in many aspects of human nature between the separate insights of psychological science and those of faith. These include the unity of body and mind, the power and perils of pride, rationality and fallibility, and the interplay of behaviour and belief. Also, science can shed light on many issues that are important for people of faith, such as human flourishing, prayer and sexual orientation.

John Wyatt contends that science alone is unable to develop a comprehensive understanding of the nature of the person. Theistic and Christian understanding is needed here for a richer and more meaningful study. The very idea of a *person* has been strongly influenced by Christian understanding. In ancient Greece, it referred to the mask that actors use, and this was developed to refer to the face someone shows to the world, the role he or she plays in society. However, the Christian claim is that humans are created in God's image, to reflect the divine character and being. It is because God's nature is personal that we are created and embodied as persons. Our humanity and the purposes of our lives only make sense in the light of our creation in God's image.

John Swinton describes David Hay's hypothesis that spirituality and therefore faith have a biological basis for humans. They are a natural part of what it is to be human, so that we possess *relational consciousness*, which makes us much closer to other people, the environment and God than we would otherwise be.

Mark Harris suggests that miracles can sometimes be described scientifically, but faith offers other non-scientific levels of significance. Thus, miracles often signify that the kingdom of God is at hand. The sea miracles illustrate the power of Jesus over nature and the forces of chaos. The feeding of the 5,000 has significance in terms of the mission of the Church, the Eucharist and the messianic banquet. Also, several stories of healing illustrate the faith of the recipient.

Tom Wright describes different ways of *knowing*, such that, for example, some sciences study the repeatable and depend on experiments, whereas history studies the unrepeatable and depends on testimony. But we do indeed know that Jerusalem fell in AD 70 or that Jesus was crucified outside Jerusalem in about AD 33 just as surely as we know the composition of a hydrogen molecule. He suggests that we need to hold together in fruitful partnership and conversation the findings of the sciences and the insights of faith, music, love, beauty, wisdom and hope. Indeed, without *meaning*, the sciences and humanities become dry and bleak.

Summary of the chapters of this book

New Atheism

Keith Ward begins by suggesting that God is a timeless being outside space and time. He is a reservoir of virtually unlimited energy, who conceives of every possible universe. This universe is brought into being by God and would not exist without God upholding it. Moreover, the statement 'God created the world' is an *axiological explanation* rather than a scientific one, since it is an explanation in terms of value rather than physical data. Keith Ward then discusses different aspects of God, such as his consciousness and creative acts, and stresses that God could function as an axiological explanation for why the cosmos exists. Asserting the existence of God is a factual claim but not a scientific one. Furthermore, belief in God can indeed be rational and based on evidence.

New Atheism is a philosophical theory about the nature of reality which has been largely discounted in philosophy as a serious contender for truth. Its core is *materialism*, which rejects personal experience, value, consciousness and purpose, and instead counts only scientific observations of physical phenomena and believes that everything that is real has to be made of matter. But the value and meaning of human life cannot be settled by scientific methods, since science is not concerned with value and meaning.

Scientific observations often suggest values that go beyond science, such as elegance, order, beauty and wonder. Although death and suffering represent a deep problem for a believer, science has helped by showing that destruction goes together with creative emergence as essential parts of the cosmic process.

New Atheists argue that science is incompatible with belief in God, but they often fail the canons of rationality in several respects, namely: not appreciating religious language and beliefs; not admitting the limits of science; not admitting the weaknesses of materialism as a philosophy; failing to distinguish between scientific and non-scientific questions; and caricaturing religious belief rather than appreciating moral and religious purpose. Instead, the existence of a rational God naturally makes the universe intelligible and ordered, and so makes science possible.

Natural law and reductionism

Eleonore Stump describes the *secularist scientific picture* (SSP) of reality, in which all can be reduced to the laws of physics. This view makes two assumptions, namely, that there is nothing to a whole other than the sum of its parts, and there is causal closure at the micro-level of physics, so that any causality at the macro-level is just a function of micro-level closure.

By contrast, according to Thomas Aquinas, human persons and human agency are instead at the centre of a discussion of natural law. For him natural law is human participation in the eternal law from the mind of God. Thus, natural law is a gift of the Creator to humans, given either by the innate light of reason or through revelation of God's mind.

For Aquinas, natural law is the law of the lawgiver, whereas for SSP it is just a description of the world at the microphysical level. For Aquinas, the ultimate foundation of reality is personal, whereas for SSP it is impersonal elementary particles.

The philosophy of SSP is *reductionist*, but that of Aquinas is anti-reductionist and neo-Aristotelian, in the sense that a thing is not just the sum of its parts but also depends on its configuration, organization or form. One example would be the function of proteins, which depends on the way the individual molecules are folded. Another would be autism, in which some psychologists suggest that the interaction between an infant and a primary caregiver plays a crucial role.

Origin and end of the universe

David Wilkinson begins by sharing the joyous excitement of Einstein when he was developing his general theory of relativity and realized that the universe is intelligible and that the intelligibility is characterized by mathematical simplicity and beauty. He stresses that an interaction between the science of the beginning and end of the universe and Christian faith is much more subtle and fruitful than would be implied by Hawking's statement that 'God is not needed at the first moment of the universe'. It is an opportunity for theology to take science seriously.

The Big Bang model describes the expansion of the universe from a time 13.8 billion years ago when it was only 10^{-43} seconds old. It is supported by observations of galaxy redshift, of the microwave background and of the abundance of helium. What happened before that time is not currently understood, but we should resist the temptation to use God to fill the gap. The Christian God is not a God who fills gaps of current ignorance, nor one who interacts with the first moment of the universe and then retires to a safe distance: rather he or she is the one who creates and sustains the laws of physics and is as much at work in the first 10^{-43} seconds as at any other time.

When the universe is 10^{12} years old, there will be no hydrogen left, stars will cease to form, and all massive stars will have become neutron stars and black holes. After 10^{14} years, small stars will have become white dwarfs, and the universe will be a cold uninteresting place composed of dead stars and black holes. Nevertheless, a Christian believes in a Creator God, which gives hope in the idea of a new creation, a new heaven and a new Earth.

Universe of wonder

Jennifer Wiseman describes the properties and evolution of galaxies, stars and planets in our universe with a sense of wonder and beauty. What has been emerging from recent advances is that evolution is not just a feature of biological systems, but is a profound frame for the entire history of our universe. Our whole universe has been changing, evolving and maturing for 13.8 billion years, transforming from energy and the first simple matter into complex galaxies and star systems, with at least one environment (the Earth) where diverse life is thriving and advanced life is contemplating its own existence.

Several thousand *exoplanets* (planets around other stars) have been detected in several ways. One is that, as a planet orbits a star, it causes the star to wobble about their common centre of mass, and the nature of the wobble enables the mass of the planet to be estimated. Another is that the light from the star decreases as a planet passes in front of it, enabling the size of the planet to be calculated. Most of the stars in our galaxy are likely to have at least one planet, and many of these planets are Earth-sized or slightly larger. In future, it will be possible to measure their atmospheric composition in order to seek biological activity.

Wiseman concludes by asking, 'Is there a purpose in the universe?' and 'Is there a purpose for our lives in the universe?' Science can inform such questions, but faith can better answer them. For her:

> the evolution of the universe from the Big Bang, through the early development of gas, stars and galaxies, through generations of stars and the production of heavy elements and complex molecules and planets, through the appearance and messy evolution of life, to complex ecosystems that include contemplative advanced life, implies a purpose of cosmic proportions.

Evolution

Kenneth Miller describes how anti-evolution movements (such as intelligent design) have fed off a perceived enmity between evolutionary science and religion, seen in statements such as 'The God of the Galápagos is careless, wasteful, indifferent' (Hull, 1991) or 'The universe we observe has precisely the properties we should expect if there is . . . no design, no purpose . . . nothing but blind pitiless indifference' (Dawkins, 1995, p. 133).

However, these are not scientific statements but faith-based assertions, and the assumption behind them is that science alone can lead us to truth regarding the purpose of existence. They have lost the sense of wonder seen in Darwin's 1889 words: 'from so simple a beginning, endless forms most beautiful and most wonderful have been, and are being, evolved.'

Most biologists agree that the capacity for life itself is built into the fabric of the natural world. People of faith should therefore respect the findings of scientific reason and develop a scientific understanding that is in harmony with their religion. Religion can in turn enlighten a scientific vision of our existence.

Kenneth Miller refers to Dobzhansky, who understood science as a way to refine and expand our understanding of the Creator's power and majesty. Also, the historical roots of modern science lie in the conviction that exploration of the natural world is an act of praise and worship.

Evolution and evil

Michael J. Murray and Jeff Schloss tackle the thorny issue of the *problem of evil*, from the point of view of both evolutionary biology and philosophy. Some have argued that the extensiveness of natural evil has been expanded by the discovery of examples of biological warfare (such as the behaviour of the ichneumon wasp) and of mass extinctions in the fossil record. However, it is not clear that evolution does increase these effects, and examples of *cooperation* rather than competition are also found.

A merciless *survival of the fittest* is seen by some as the central element of diversity and evolutionary change. But cooperation can also drive evolution, and natural selection does not necessarily involve competition, since evolutionary change can also occur by the opening up of new habitats. For others, evolution precludes the existence of progress and of natural goodness. However, many aspects of evolution do exhibit progress as positive traits develop over time, and major evolutionary transitions often involve emergent levels of enhanced functionality.

Murray and Schloss then move on to address the question of *why God allows evil*. Are there examples of suffering that an omnipotent and omniscient being could have prevented without losing a greater good or permitting a worse evil? Is there an explanation of evil that is true for all we know? They consider critically the following options that have been proposed:

- *the fall* – the wrongdoing of Adam and Eve had consequences for the relation with God and the universe;
- *freedom* – evolution has the freedom to lead to good or evil;

- *evolution* – evil is created by the impersonal process of evolution rather than by God;
- *neo-Cartesianism* – the consciousness of animals is quite different from human consciousness, such that they may have no conscious awareness of pain, or pain may not bother them;
- *embodied intentionality* – pain is a necessary way to prevent animals injuring themselves;
- *chaos-to-order* – natural evil is a by-product of a universe that proceeds from chaos to order by law-like regularity.

Genes

Pauline Rudd stresses that the body is a complex entity without a simple hierarchy, made up of many macro- and microstructures with feedback loops that enable us to survive in a changing world. We are made up of thousands of dynamic systems, many outside our conscious control. Our 30,000 genes are not a blueprint for our bodies, but they do contain information that represents potential and imposes constraints.

Genes are DNA-based units that exert effects on an organism through RNA or protein products. They possess four bases assembled on very long molecules of deoxyribonucleic acid (DNA). One strand of DNA with the bases attached and its complementary partner combine to form a double helix chromosome structure. Humans possess 23 tightly folded pairs of chromosomes.

Genes can replicate themselves. Many genes code for proteins with multiple functions. Genes respond to signals from their environment telling them that the protein is needed. Genes can be altered or mutated, perhaps by miscopying, which may produce a protein with a useful or destructive function. Genetic diversity makes each of us unique and helps ensure the survival of our species.

Science does not deal with certainty but with *knowledge and probability*, so that the natural world works by making new beginnings that arise from uncertainty. At a biochemical level, we are certainly more than our genes, for it is the external and internal environments that trigger gene expression. Our decisions are guided by the possibilities and limitations imposed by our genes and also by the environment. But living a fulfilled life means being integrated in a complex world where we find a niche to flourish physically and emotionally.

Psychology and faith

David Myers discusses how he reconciles being a psychologist and a person of faith. He has an open-minded attitude that has caused him to change many of his earlier beliefs, so that he now believes that:

- parents have modest effects on their children's personalities and intelligence;
- electroconvulsive therapy can often relieve depression;
- the unconscious mind dwarfs the conscious mind;
- traumatic experiences are rarely repressed;
- sexual orientation is a natural, enduring disposition rather than a moral choice.

Psychology and faith intersect in the following topics:

1 the effect of values and assumptions on psychology;
2 application of psychological insights to religious communities;
3 the psychology of religion;
4 psychological and religious understanding of human nature;
5 the observed effects of religion;
6 tensions between psychology and religion.

There are parallels between psychology and Christianity when discussing the unity of mind and body, pride, rationality and fallibility, as well as behaviour and belief. Furthermore, people with a religious faith show greater generosity with time and money, live longer and are happier. Myers also discusses the effect of prayer in living as God's people.

Finally, he puts forward a strong case for same-sex marriage, on the grounds that:

- all humans have a deep need to belong;
- marriage contributes to flourishing lives;
- individualism is corroding marriage;
- sexual orientation is a natural disposition rather than a choice;
- sexual orientation is also an enduring disposition that is seldom reversed by willpower or therapy;
- the Bible has nothing explicitly to say about enduring sexual orientation or about loving, long-term same-sex partners.

Nature of the person

John Wyatt considers what it means *to be a person* from his perspective as a neonatologist working with premature babies. Some secular philosophers suggest that newborn babies and people with severe dementia or learning disabilities should not be regarded as persons and so should have fewer rights and privileges. Personhood in their view is determined by high-level cognitive functioning, including having preferences about continued life and interacting in a sophisticated way with others.

Wyatt also discusses the nature of consciousness and the philosophical perspective of non-reductive *physicalism*, whereby the brain is entirely physical and material in nature, but mental states can emerge from physical neuronal processes and react back down on neuronal activity.

He contrasts this with a Christian understanding of the nature of person, in which reality consists of the personal as well as matter and energy, so that persons are not reducible to or limited by matter and energy. A person is a different kind of reality, namely, one that knows and is rational, communicative, creative, moral and loving. A person is a profound unity, with both a physical, material aspect and a personal, immaterial aspect.

Indeed, God's ultimate being is described in terms of three persons giving themselves to one another in love. The Christian claim is that we are made in God's image, created to reflect the divine character and being. Each human person is unique, made for relationships with others. An alternative version to '*Cogito ergo sum*, I think therefore I am' could be 'You love me, therefore I am'. Thus, in his career, John Wyatt has been called to recognize newborn babies as mysterious others to whom he owes a duty of care and protection.

Science, spirituality and health

John Swinton lays out the issues regarding ways in which science and religion can come together in a mutual quest for *health* and human well-being. He starts by highlighting the differences between *spirituality* and *religion*. Spirituality is an individualistic experience that all people have and that includes a quest for meaning, purpose, hope, value and love. Religion includes practices and beliefs about

God that have roots, traditions and philosophical underpinning and that assume spirituality comes from outside.

He gives an account of scientific results that indicate religion may have many health benefits, suggesting possible causes and describing both the value and limitations of such research. He also contrasts a medical and Christian definition of health.

Important research by David Hay is outlined, proposing that human beings are *innately spiritual*, that they are hardwired for spiritual experience. This natural spirituality is revealed in the way that children have an intrinsic sense of wonder, awe, relationality, connectedness and acceptance of things beyond their understanding. Human beings are therefore by nature interconnected and relational rather than being autonomous and individualistic. Hay describes this as *relational consciousness*, which is experienced when one realizes one's interconnectivity with others, with God and with the world.

If Hay is correct, there are huge consequences for Western secular society. Rather than religion being, according to Freud or Marx, a social and psychological construct designed to alleviate anxiety, it is Western individualism and secularism that are socially constructed. This helps to explain why, even though traditional religions are declining in popularity, many people are still searching for the spiritual. Also, rising rates of loneliness, depression and anxiety can be understood if our natural biological desires for spirituality are being suppressed by the current cultural emphasis on individuality rather than relationship.

Swinton concludes by suggesting that a way forward to enhance the relationship between religion and science in healthcare is through the concept of *hospitality*, in which the integrity of each side is valued, respected and honoured by the other.

Miracles

Mark Harris considers the miracles of Jesus and suggests that the modern worldview that miracles are scientifically impossible and that belief in them is a relic of a bygone primitive age is unsustainable in the light of the complexity of the relationship between miracle and science.

David Hume defined miracle as 'a transgression of a law of nature by a particular volition of the Deity'. However, this regards nature as a rigid closed system, which is unreasonable in the light of modern developments of quantum mechanics, complexity and emergence. Furthermore, the miracles of the exodus may be regarded as normal events that do not violate nature, but are still miracles of timing.

Jesus was known as a miracle worker, but his miracles are very diverse, some of them involving healings and others involving nature, such as the stilling of the storm. Do the miracles of Jesus contradict science? Perhaps sometimes no (e.g. the miraculous catch of fish) and perhaps sometimes yes (e.g. the raising of Lazarus). In the Gospels, the miracles, however, often have a deeper significance, suggesting that Jesus has power over nature and 'the kingdom of God is at hand'.

Trusting the New Testament

Tom Wright asks whether a scientist can trust the New Testament. He suggests that the great lie of today's scientism is that science has proved Epicureanism (that the world works by itself since the gods are far away), but he stresses that science cannot adjudicate between different philosophical positions.

He compares different *forms of knowing*, in particular in science, in history and in the worlds of religion, culture and art. Then, he asks whether we can take the story of Jesus seriously as history, and in particular discusses whether we can believe in the resurrection, believe in miracles and trust the record of Jesus.

His conclusion is that we can indeed trust the New Testament to tell us about new creation, and about a power that generates new modes of knowing. This trusting is not a cool detached cerebral activity but involves opening ourselves as participants in, rather than spectators of, the source of life.

2

God, science and the New Atheism

KEITH WARD

What is meant by 'God'?

The word 'God' has a fairly well-defined dictionary definition. *The New Shorter Oxford English Dictionary* puts as its first definition, 'A superhuman person regarded as having power over nature and human fortunes'. It also offers: 'An adored or worshipped object; something exercising great or supreme influence'. These definitions are broad enough to cover ancient Greek gods like Zeus and the various deities of many tribal peoples, as well as the one God worshipped by Jews, Christians, Muslims and others. Strict monotheism, the belief that there exists only one God, and that God is the creator of nature and in fact of everything other than itself, seems to have arisen among the early Hebrew tribes, and was first clearly formulated in the later Prophets of the Hebrew Bible (the Old Testament).

We know what the dictionary means by speaking of God as a 'superhuman person', but most Christians would feel that it falls far short of being a satisfactory definition. In some ways it is positively misleading. It sounds more like a definition of Superman than a definition of the creator of heaven and Earth, and of the whole of space and time. So the first step in finding a more satisfactory definition is to reflect on what it means to say that there is a creator of space and time.

For a start, it means that the Creator is beyond our space and time, because whatever creates space cannot be in that space, and whatever creates time cannot be in that time. So God exists beyond our space–time. That does not sound like a 'superhuman person', because even superhuman persons would be in space and time. They would be like us, but rather more powerful, and that is just not good enough for a Creator God.

If we ponder a little more, it becomes clear that a being that is not in our space–time is really going to be very unlike any sort of human person. When we say that 'God created the universe' we are not saying that, just before the universe began, God existed all alone, and wondered whether or not to create a universe, and what sort of universe it would be. Then after some thought God decided to create this universe. Then God did create it, and the universe began. So God existed before the universe did, and then (at some specific time) decided to create it, after which the universe existed as well as God.

As David Wilkinson indicates in his chapter in this book, the great Christian theologian Augustine of Hippo pointed out that there is something very wrong with such a story. What is wrong is this: if God created our time, then there obviously was no time like ours in which God existed before God created it. If our time began, then there was no time of this sort before it. And if there was no time before it, then God did not exist before the universe did.

This might sound shocking, but it is what the great Christian theologians have said. God, they say, does not exist in time at all, and so God did not exist *before* the beginning of the universe. That does not mean that God does not exist at all. It means that God is a timeless being (the word usually used is 'eternal') who generates the whole of time, and upon whom all time depends, so that there would be no time without God.

If this sounds strange, it is just what many modern physicists say when they say that this universe, and maybe many other universes too, all originate from a realm beyond space–time. That realm they often call the *quantum vacuum*, but it is not just nothing at all. It is a realm of vast potential energies and beautiful and intelligible quantum laws. It is eternal and maybe it is necessarily what it is, and it is the cause of every space–time universe that comes into existence.

Of course God is not a quantum vacuum. But God is a being of vast power (enough power to create entire universes) and great beauty and intelligibility. God is eternal and maybe necessarily what God is, and is the cause of every universe that comes into existence. The main difference from the quantum vacuum is that God *knows* every possible universe that could ever exist, and *chooses* this universe for a good reason. Whereas the quantum vacuum, as conceived by most physicists, would be unconscious and would have no purpose, God is conscious or aware, and has a purpose in creating, bringing into existence, a universe.

41

This is a pretty big difference, so big that Stephen Hawking has apparently said that God is not needed for the creation of the universe (see David Wilkinson's comments about it in his chapter). I do not think this opinion is correct, and I think that Professor Hawking is thinking of a very naive idea of God as a superhuman person. But what I find interesting is that modern physicists seemingly have no problem with thinking that our universe depends upon a timeless reality beyond it. So they should have no problem with the idea of God as an eternal creator of the universe. And their thinking helps to make it clear that God is not just a 'superhuman person'. God is a reservoir of virtually unlimited energy, a being who conceives of every possible universe there could ever be, and a being whose purpose is concerned with billions of galaxies and stars, not just with human beings on a rather small and isolated planet circling round a rather small star in a rather ordinary galaxy. And God does not just have 'power over nature and human fortunes'. God brings whole universes into being, and nothing at all would exist without the being of God upholding it. This is the God that Christians worship with awe, both for the unthinkable power of the Universe-Creator, and for the incredible fact that this Creator has encountered them in the person of Jesus with unlimited love.

Scientific explanation

I have said that the Christian God is the Creator of the universe. This can superficially sound like an attempt to give a rather weak scientific explanation. If you say, 'God created the world', that sounds like a scientific explanation. It explains why the world is the way it is, and why it exists, by positing a hidden cause, God. But the surface grammar is misleading. Scientific and religious explanations are different in kind.

Scientific explanations generally refer to physical data that are in principle observable and publicly testable. The explanations can ideally be formulated in mathematical terms, or at least the data they deal with can be measured and quantified with precision. Experiments can be devised which test scientific proposals, and enable us to predict and sometimes use physical processes for improving the quality of human lives. Scientific explanations are tremendously useful, and scientific methods of formulating *laws of nature* and of repeated experimental observation are essential to a modern understanding of the world.

But not all explanations are scientific explanations. 'God created the world' does explain why the world is the way it is, at least in part. But 'God created the world' does not give an ordinary causal explanation of some hidden physical reality that preceded this universe in time, that we could experimentally test, describe in a neat mathematical equation, and perhaps use to create improved universes in future. So what sort of explanation is it?

Axiological explanation

It is a sort of explanation that is perfectly familiar to us in everyday life, and one name for it is *axiological explanation*. It is what we use when we try to explain, for example, why rational people act as they do – rational people act in order to obtain something they value. So axiological explanation is the explanation of a process in terms of value. It explains why people act as they do in terms of valued states that people want. It has four major elements. First is the identification of some state or process as of intrinsic value, as being worth choosing for its own sake alone. This entails the second element, which is awareness of a range of alternative states on the basis of which such an evaluation could be made. Third is the assumption that a choice can be made. And fourth is the conscious appreciation and enjoyment of the value, without which all values would remain merely hypothetical rather than actual. Such explanation presupposes that intrinsic values do exist, that there is consciousness of both their possibility and actuality, that purposive choices can be made (choices made for the sake of realizing a specific value), and that there are feelings or desires that can in principle be satisfied.

Axiological explanations are not usually used in the natural sciences. Strictly physical sciences do not ask whether anything is of intrinsic value; they set aside questions of consciousness and of subjective feelings, and they are extremely wary of speaking of purposes or goals in natural processes. The human sciences, like some forms of psychology and economics, may introduce such topics, but they usually retain a primary interest in recording publicly observable behaviour, in collecting data that can be measured in some way, and in attempting to frame significant generalizations that can be tested in varying contexts. They are usually content to record trends and correlations rather than to frame precise *unbreakable* laws, and they are usually keenly aware of the many exceptions and unique cases that will qualify their general conclusions.

To give an axiological explanation of the whole universe would be to identify the intrinsic values that it realizes, to suppose that the cosmos is selected from a number of alternatives precisely because it realizes those values, and therefore to postulate that there is a consciousness – *God* – who envisages, selects and appreciates those values. This could not, as in the human case, be a matter of recording the publicly observable behaviour of such a trans-cosmic consciousness, or of measuring its behaviour, or of framing testable generalizations about it that would apply to all gods of the same sort, at least not if there is in principle only one God.

Aspects of God

God is thought by Christians to be a unique case, and is not a physically observable object, so it is hard to see how any physical descriptions or scientific generalizations could be offered in the case of God. This means there could be no scientific explanation of God's actions. Nevertheless, God could function as an axiological explanation of why the cosmos exists as it does, namely, for the sake of the values that it realizes and that God, and also other agents, can enjoy.

God's consciousness is utterly inaccessible to humans, since God has no locatable physical body to express divine thoughts and feelings. Moreover, it is a consciousness that is not dependent on some complex physical structure like a brain, so it has a sort of causal priority over matter that is quite unfamiliar to us. God does not know things, as we do, through sense organs. God's knowledge is direct and unmediated, and it will cover not only the whole universe, but also all the alternative universes that could possibly exist.

Moreover, God's desires and acts will not be whimsical or arbitrary. God will discern the true nature of all intrinsic values, and God's creative acts will be governed by that discernment. Thus, for most theologians, as for Plato and Aristotle, the being of God will itself be of supreme intrinsic value, since it contemplates all possible values without change, frustration or decay. God is the supremely Good and Beautiful, and that is, from an axiological viewpoint, the best of all reasons for the existence of anything.

God's agency would be the source and origin of the whole cosmos. As such, it would be beyond space and time, as their origin. Its knowledge and agency would thus be vastly different from ours. The Supreme Good that cannot fail to be, that is self-existent and perfect in actuality, is as far superior to human consciousness and personality as our consciousness is to that of a beetle. There is no hope that the methods of physical science could ever be used successfully to provide a scientific explanation of God or of God's actions.

Belief in God

Yet to assert the existence of such a God is certainly to make a factual claim, a claim about how things are. God is the spiritual creator of the physical universe. But this is not a scientific claim. It does not offer any particular physical explanation of how the universe came into being, and it does not offer publicly verifiable and experimentally testable evidence for the existence of God.

However, it would be quite wrong to say that it is irrational, or that it is based on no evidence. Belief in God is rational, because it is based on our knowledge that consciousness and intentional agency are fundamental features of reality. Many, indeed most, classical philosophers have argued that consciousness, and not unthinking matter, is likely to form the basic causal structure of reality. Belief in God is based on evidence, the evidence of personal conscious experience, of experience of value, especially in morality and art, and experience, common in many religious traditions, of liberation from egoism and conscious unity with a supreme Good.

Not all good evidence is public or experimentally testable. We all know our private thoughts and feelings in ways no one else can. As for experimental tests, it would actually be immoral to devise experimental tests for whether people we know really love us. The deepest personal relationships depend upon commitment and trust, upon the cultivation of a rich inner complex of thoughts and feelings that we can never fully express, and upon loyalties that go beyond what we could strictly demonstrate to be the case.

Ironically, logical positivism, the philosophy that made verification by the senses a condition of making meaningful factual assertions, was unable to establish even the existence of a public world of physical objects, since it was unable to prove that any public world even existed, as that would assume a set of other minds that were not directly verifiable by the senses.

Verification of some sort is important. But why should it be limited to sense experience, and why should anyone insist that verification has to be conclusive and beyond dispute, in a world as transient and ambiguous as this? Intimations of transcendence and of value are sorts of verification. Science does not deal with them, but there is no reason for science to deny them.

The New Atheism

Why, then, should there have arisen in the last few years a group of writers, usually with no great interest in and little respect for philosophy, who are resurrecting the rather old and historically exploded legend about a war between science and religion? I think it is mainly because of a rejection of personal experience as a reliable source of knowledge, and the consequent downgrading of value, consciousness and purpose to being subjective by-products of a wholly material reality, of which science gives the only reliable form of knowledge.

This is not in fact a scientific theory. It is a philosophical theory about the true nature of reality, a theory which is presently very fashionable, but has historically been largely discounted as a serious contender for truth. It is extremely odd to despise philosophy and yet to rely on such a very complicated and highly disputed phil-osophy as materialism. To say that the whole of conscious experience, with its rich and value-laden content, is either reducible to physical processes in the brain or is wholly causally dependent on such pro-cesses is a hypothesis that is far from being established scientifically, so no view which purports to rest only on the well-established findings of science should assume it to be true. Such a hypothesis rests on a commitment to philosophical materialism, a system of thought that seems to many philosophers to undermine the very basis of human knowledge, which in the end lies in conscious experience.

Materialism is in fact self-contradictory if it asserts as true the proposition that 'only public observations of physical phenomena in space and time can count as evidence for true beliefs', since the evidence for the truth of this proposition cannot be any set of public observations.

It will not do to say that the proposition is not a truth, but simply a declaration that one will not count anything but public observation as evidence. If such a declaration is to be reasonable rather than quite arbitrary, it must be based on something like the consideration that only public observations provide useful or fruitful knowledge. But that begs the main question at issue: are our experiences of value and transcendence, our struggles to understand our own lives and learn how to live well, all useless and fruitless? Are our often agonized attempts to find meaning in our lives, to face up to the anguish of despair and death, to find something worthwhile in our inner struggles, to be consigned to being pointless by-products of unconscious material processes?

Perhaps here we touch the real heart of the New Atheism – a rather old atheism, in fact, that reached its zenith with Nietzsche and Marx. For this is not just an abstract philosophical debate between idealism (the philosophical view that something mind-like is the basis of reality) and materialism (the philosophical view that everything that is real has to be made of matter). It is not a debate between religion and science at all. It is a passionate debate about the value and meaning of human life and experience. Such debates cannot be settled by scientific methods, which are not, as such, concerned with questions of value and meaning.

Probably the best-known 'New Atheist', Professor Richard Dawkins, in his book *The God Delusion*, has what he takes to be a knock-down argument against God. But the argument is very weak. It is that all intelligent consciousness has to be the end result of a long process of evolution (as it is in humans). Complex beings have to evolve from simple beginnings, because the simple is more likely to exist than the complex. But God, a creative mind, would have to be more complex than the universe, and so God is less probable than a universe without God, and needs at least as much explanation as the universe does. Therefore God is a useless hypothesis.

There are three main problems that show why this argument does not work. The first problem is the assertion that all intelligent consciousness has to be the result of a long evolutionary process. There is no reason why this should be the case. The idea of a consciousness that is aware of every possible universe, that can bring universes into being intentionally, and that exists without being in a space–time manifold, is a perfectly coherent notion. It is sheer dogma to deny that possibility, a possibility that most classical philosophers have usually embraced.

The second problem is this. Dawkins holds that it is more probable that a universe of simple elements exists than that an internally complex mind exists. The fact is, however, that the notion of probability does not even make sense when there is nothing already in existence to give rise to probabilities. For instance, the probability of throwing a six with a die is 1 in 6. We know that because we know there are just six sides, and we assume that the die is not weighted. But if you do not know at all what sorts of things might exist, and you ask which is more probable, a God or a universe without God, you can have no idea how many possibilities there are, or whether they are weighted in any way. When you do not know those things, you simply cannot assign a probability at all. The notion of probability does not make sense. God is not less probable than the universe without God.

The third problem is about the idea of God needing explanation. It is true that God is internally complex, in having huge numbers of ideas of possible universes, for example. But God is not made up of many simple parts stuck together. God is one consciousness, just one being, who explains the universe by affirming that the universe exists to realize a valuable purpose. That is a very simple and economical explanation. Does God need explaining just as much as the universe does? No. There can be no scientific explanation for God, because God is not a physical object, and because God is beyond time, and therefore cannot possibly be caused. Being caused entails first of all that God was not, and then (at a later time) God was brought into being, but an eternal being cannot possibly ever have been brought into being. God is, by definition, a being who cannot be caused by anything.

Yet God does have a sort of explanation – what I have called an axiological explanation. That is, God is supremely worthwhile. That is the best sort of reason there could be for anything. There could, of course, be such an explanation for the universe, and it would be a good explanation. However, the universe is not supremely worthwhile. Christians believe that the created universe is good, but they do not believe that it is all supremely worthwhile. Its goodness is often frustrated by evil, and remains largely potential, something we hope will be realized in the future. It is only God who is supremely worthwhile, and that is why Christians worship only God. They should respect and care for the universe because it is God's creation, but they do not worship it.

God therefore does not need, and could not possibly have, the same sort of explanation that the universe needs. But God has a good explanation of a different sort. Any scientific explanation of the universe will end with the idea of some, allegedly simple, ultimate laws of nature and some allegedly simple physical states (it is very doubtful, by the way, that these things really are very simple). Those things exist without consciousness or purpose or value; they are just there for no reason. That is not a very good explanation after all. God provides a much better explanation, namely, that the divine being exists because it is of supreme goodness, which is the best reason there is for existing. God cannot have a scientific explanation, but God gives a good explanation for why the ultimate laws of nature are as they are – because they produce a universe with a good and valuable purpose. Of course no human fully understands the goodness and purpose of God, but Christians believe that Jesus has revealed that this goodness and purpose is self-giving and self-fulfilling love. That is enough for humans to see that God is not superfluous, but is the foundation and the goal of all that is of value in the world, and that God has given humans the purpose of loving and enjoying God for ever.

Going beyond science

However, there is a problem with the Christian view too. Does science not show that nature is cruel, purposeless and pointless? No, as Kenneth Miller shows in his chapter in this book, that is a value judgement and not a conclusion of scientific study. Of course, if we believe that the cosmos has a purpose – to produce distinctive sorts of value – then examination of the cosmos is relevant to whether there are such values, whether it is reasonable to see the cosmos as directed to producing them, and whether it is such that an intelligent consciousness could have created it. But we have to go beyond science to answer such questions. We have to engage in philosophy, asking what sorts of values there might be, how and in what way they might exist, and how they might connect with various sorts of purpose.

Scientific observation of the cosmos suggests some values very strongly: the elegance and ordered complexity of the laws of nature, the beauty of the galaxies, the creative emergence apparent in the majestic processes of cosmic evolution, the incredible integration of simple parts into complex organized wholes, the development of understanding and appreciation in 3 pounds of grey matter in the human skull. Science is not an emotionless discipline, and most scientists are inspired with amazement and awe by the sheer grandeur of the universe.

Yet such evaluations and emotions are not parts or conclusions of any strictly scientific theory. They might motivate scientists, and they may be evoked by scientific studies, but they do not occur *in* scientific theories. Good scientists may even fail to have them. Steven Weinberg famously commented that the more he understood the universe, the more pointless the universe seemed. Yet the understanding and appreciation of the complexity, magnificence and rich variety of the physical universe that science can bring is precisely one of the things that might give the universe a point or intrinsic value. Questions of value and purpose, of the place of consciousness in the universe, of the moral importance of human persons, and indeed of the importance and status of science itself, must pay close attention to scientific data, but science does not provide conclusive answers to them.

A fundamental element of belief in God is that there is intrinsic and objective value in such things as beauty, intellectual understanding, creativity, and compassionate and cooperative personal relationships. For a theist, those values are instantiated supremely in God, and the universe expresses some aspects, images or reflections of them, insofar as they can be embedded in time. Human fulfilment consists in shaping human awareness to appreciate them more fully, to celebrate them and to create new temporal expressions of them. This is what gives human existence its purpose.

Death and suffering

There is much in the universe as scientific observation discloses it that tends to support such a religious view. But there are undoubtedly problems too. The hardest problem for any theist is to account for the existence of death and suffering in the cosmos if it were created by a benevolent God. This is not a new scientific problem, but an old philosophical problem of rational consistency.

The sciences do, however, adduce some relevant facts. One of the most significant is the discovery that destruction and suffering seem to be essential and ineliminable parts of the cosmic process. Without the destruction of stars, heavy atoms would not form. Without the law of entropy or universal long-term decay, the temporal process would have no direction. Without the competition of species for survival, the selective effects of evolution would not occur.

The emergent properties of the cosmos come about through a sort of creative exploration of possibilities that inevitably involves failures as well as successes. In the light of much modern science, it becomes plausible to say, as Steven Weinberg (1977) does allow, that humans, as the emergent carbon-based life forms we are, could not exist in any other universe than this, with its laws of gravitational attraction, electromagnetism, strong and weak nuclear forces, and entropy that entail destruction as well as creative emergence throughout the universe. God might have created another universe, but it would not have us in it. So if God wants us to exist, with the distinctive values we can realize, this is the universe there has to be. This is not a scientific remark, but perception of the universal interconnectedness and destructive–creative polarity of the universe derives from a plausible interpretation of modern science.

In this way, discoveries about the nature of the universe may affect our conception of a personal creator. It has, I think, become implausible to think of God directly intending every part of this universe to be as it is, since much in the universe is either destructive or random (not fully determined). But it remains plausible to think that God has created the laws and processes of the universe for the sake of the distinctive sorts of value the universe will produce. God sets up basic structures in the cosmos that will guarantee the achievement of a desired goal, but also allows enough indeterminism within those structures for intelligent creatures, when they evolve, to make reasoned choices between alternative futures. It is plausible to think that the ideal goal that exists in the mind of God will have some specific causal influence on the physical processes of the universe. We may find it difficult to conceive of how such influence will be felt, since we lack a theoretical model that is adequate to it. But, if we have made the initial postulate of God, the observed facts seem compatible with a view that sees God not as determining every event, and not as interfering occasionally in a closed and complete physical system, but as exercising a general attractive or teleological influence, which may be felt as a propensity to life, consciousness and intelligence in an open and emergent universe. This will be more apparent in some crucial instances than in others. God's influence on the world might be real, and yet limited by many other causal factors that are necessary conditions for the existence of carbon-based intelligent beings.

Is science incompatible with belief in God?

The so-called 'New Atheists' argue that acceptance of science is incompatible with belief in God. They claim to have a completely rational, indeed the only rational, view of the world and human life. But there are five important respects in which these writers often fail to be completely rational:

1 They often tend to caricature religious practices and beliefs, and to discuss them only in their most naive and extreme forms. They thus neglect the very first principle of critical thinking, which is to state one's opponents' views as fully and fairly as possible.
2 They do not admit that there are limits to scientific enquiry, and that there are many factual questions – such as moral or philosophical questions – which fall outside any such theoretical framework.
3 They do not see or admit the philosophical weaknesses of materialism as a philosophical theory, and the strength of more theistic or idealist views, which have been almost universally espoused by both Western and Eastern philosophers.
4 They fail to draw an important distinction between the well-attested findings of natural science and wider worldviews of a philosophical nature, like materialism, that remain undetermined by science.
5 They often seem to have a deeply emotional antipathy to the idea of a moral and spiritual purpose for human life, which antipathy is rooted in a view of religion as anthropomorphic, literalistic, and opposed to life, joy and freedom. To characterize all religion in this way is to fail to make important discriminations between various kinds of belief in God.

Belief or disbelief in God, like all beliefs entailing definite practical commitments, can be a highly emotional matter. But there is a place for reason in considering such beliefs. It is ironic that those New Atheists who like to place themselves under the banner of reason themselves break some of the basic rules of rational discussion. Even worse, they espouse a worldview that makes reason an accidental by-product of a long and pointless struggle for survival. On such a view, it is hard to see why reason should be regarded as a reliable path to truth. If, however, you believe in a God who, as the first verse of John's Gospel states, created the world through reason (*logos*), then you would expect the universe to be as intelligible and ordered as it apparently is. The final irony is that it is belief in a rational God that makes science possible, whereas in an atheistic universe it is a complete surprise that there is any rational structure to the universe, or that human reason can make any sense of it. Far from there being a war between science and religion, it seems that belief in a rational and supremely valuable God is an important support for good science.

3

Natural law, reductionism and the Creator

ELEONORE STUMP

Natural law as the laws of physics

Trying to summarize the view of the world generated by the secularist appropriation of science now common in Western culture, Simon Blackburn describes things this way:

> the cosmos is some fifteen billion years old, almost unimaginably huge, and governed by natural laws that will compel its extinction in some billions more years, although long before that the Earth and the solar system will have been destroyed by the heat death of the Sun. Human beings occupy an infinitesimally small fraction of space and time, on the edge of one galaxy among a hundred thousand million or so galaxies. We evolved only because of a number of cosmic accidents ... Nature shows us no particular favours: we get parasites and diseases and we die, and we are not all that nice to each other. True, we are moderately clever, but our efforts to use our intelligence ... quite often backfire ... That, more or less, is the scientific picture of the world.
>
> (Blackburn, 2002, p. 29)

I will call a view such as this 'the secularist scientific picture' (SSP, for short) to distinguish it from a mere summary of contemporary scientific data. It remains a widely held picture of the world, even though, as I will show in what follows, research in various areas is making inroads against some parts of this view.

In SSP, as I will understand it for the purposes of this chapter, the natural laws Blackburn refers to are typically taken to be the laws of physics, and all other laws are supposed to be reducible to the natural laws of physics. All *things* in the world are thought to be reducible to the fundamental units of matter postulated by physics and governed by the natural laws of physics.

One important presupposition of SSP is a metaphysical rather than a scientific principle, namely, that a thing made of parts is identical to the parts that are its constituents. On this view, there is nothing to a whole other than the sum of its parts. And, of course, the same holds for each of the parts. Each part is also nothing more than the sum of *its* parts, and so on down to the most fundamental level. Ultimately, everything is identical to the most fundamental parts that constitute it. In SSP, these are the elementary particles governed by the natural laws of physics.

The metaphysics incorporating the principle that constitution is identity is one version of reductionism. As Robin Findlay Hendry puts it, 'the reductionist slogan is that x is reducible to y just in case x is "nothing but" its reduction base, y' (Hendry, 2010, p. 209). Applied to theories rather than things, reductionism holds that all the sciences reduce to physics, and all laws are reducible to the laws of physics, together with bridge laws connecting theories in the higher-level sciences to theories in physics.

Reductionism is often thought to rest on another metaphysical claim as well, namely, that there is causal closure at the level of physics. Apart from quantum indeterminacy, there is a complete causal story to be told about everything that happens; and that complete causal story takes place at the level of the elementary particles described by physics. On the view of natural laws in SSP, then, any causality found at the macro-level is just a function of the causality at the micro-level of physics. Because there is causal closure at the lowest level, the causal interactions among the fundamental particles of a thing are not open to interference by anything which is not itself at the most fundamental level and governed by the natural laws operating on that level. And everything that happens at any higher level, from the chemical to the psychological, happens as it does just because of the causal interactions among the fundamental physical particles involved.

So, for example, any act of a human being is explained by events at the level of bodily organs and tissues; these are explained by events at the level of cells; these are explained by events at the level of molecules; these are explained by events at the level of atoms – and so on down to the lowest level, at which there are the causal interactions among the elementary particles postulated by physics and governed by the natural laws of physics. The causal interactions of things at this lowest level thus account for everything else that happens, including those things human beings do.

Or, to put the point of this example in a more provocative way, in SSP love and fidelity, creativity, the very achievements of science, and any other thing that makes human life admirable or desirable is itself just the result of the causal interactions of elementary particles in accordance with the natural laws of physics.

For many people, me included, the implications of SSP seem highly counter-intuitive. Can the laws of all the other sciences really be reduced to the laws of physics? Is everything really completely determined by causal interactions at the microphysical level? Could it really be the case that the mental states of a person are causally inert as far as his or her own actions are concerned? Could an act of will really be both free and yet also causally determined?

Natural law in the thought of Thomas Aquinas

It is instructive to reflect on SSP by contrasting it with the very different view of the world held by the medieval philosopher Thomas Aquinas. Aquinas talks of natural law, too; but the notion of natural law in the thought of Aquinas is nothing like the notion of natural law in SSP. With respect to the notion of the natural law in Aquinas's thought, human persons and human agency are not rendered marginal or even invisible, as they seem to be in SSP. They are at the centre of the discussion. (For more discussion of Aquinas's notion of natural law and its place within his meta-ethics and normative ethics, see the chapter on goodness in Stump (2003).)

When Aquinas explains his notion of natural law, he says that the *natural* law is a participation on the part of a human person in the *eternal* law in the mind of God (*Summa Theologiae* IaIIae 91.2). And, when he explains the *eternal* law, he says that it is the ordering of all created things as that ordering is determined in the mind and will of the Creator (*Summa Theologiae* IaIIae 91.10). For a created person to participate in the eternal law of God, then, is for that person to have a mind and will which reflect their origin in the Creator: the natural law in created human persons is an analogue of the eternal law in the Creator.

So one way to understand Aquinas's account of natural law is as a gift of the Creator to the human persons he has created. It consists of a pair of habits, one in the will and one in the intellect, which is given to human beings either by means of the innate light of reason or through the Creator's revelation of God's own mind to God's creatures. Although, apart from revelation, these gifts are implanted innately, they are so far in the control of the creature that a person's exercise of his or her free will in evil acts can corrupt them. Nothing about God's rendering the natural law innate in human persons takes away from them their free agency.

Just as many people find the implications of SSP counter-intuitive, so, for many people, the implications of Aquinas's account of natural law, grounded as it is in his metaphysics and theology, seem counter-intuitive too. Can everything in the world really be traced back to an omnipotent, omniscient, perfectly good Creator? Could it really be the case that a human person has the causal powers of intellect and will which reflect the eternal law in the mind of the Creator? Or, to put the question in a less theological way, could the action of something at the macro-level, such as a human being, exercise causality, from the top down, as it were, without being itself determined at the micro-level?

Double vision

Any attempt to hold in one view the very different notions of natural law in SSP and in the outlook of Aquinas can induce vertigo. How is one to understand the differences in worldview between the two, and how could one even begin to adjudicate their competing claims?

It will be profitable to begin by considering their highly varied foundational metaphysics.

As has often been remarked, one notable difference between the notion of natural law in SSP and the Thomistic notion of natural law is that, for Aquinas but not for SSP, natural law is the law of a lawgiver, whose mind is the source of the law and whose relation to and care for other persons lead him to promulgate the law. On the view of natural law in SSP, the whole notion of law is only metaphorical or analogous. A natural law of physics understood as SSP sees it is just a generalization describing the nature of the world at the microphysical level.

This dissimilarity is correlated with a much greater difference as regards the ultimate foundation of reality. In SSP, the ultimate foundation of reality consists in those elementary particles described by the ultimately correct version of contemporary physics and their causal interactions governed by the natural laws of that physics. All the sciences are reducible to physics. And everything that there is is reducible to the elementary particles composing it. Persons are no exception to this claim. Persons too are reducible to the elementary particles that constitute them. At the ultimate foundation of all reality, therefore, there is only the non-personal.

What is challenging for SSP therefore is the construction of the personal out of the impersonal. The mental states of persons, their free agency, their relations with each other all have to be understood somehow as built out of the physically determined interaction of the non-personal.

On Aquinas's view, things are in a sense exactly the other way round. That is because for Aquinas the ultimate foundation of reality is God the Creator. On the Thomistic worldview, the ultimate foundation of reality is therefore precisely persons.

It would not be hard, I think, to trace the notable differences between SSP and Aquinas's worldview, as implied by their varying notions of natural law, back to the great dissimilarity in their metaphysical views regarding the ultimate foundation of reality. But, given this radical difference between SSP and the Thomistic worldview as regards such foundational matters, is it so much as possible to reason about their competing claims?

Even if the recent history of philosophy did not make us pessimistic about the prospects for success when it comes to arguing over the existence of God, it is clear that it would not be profitable in a short chapter to tackle a disagreement of this magnitude head-on. It is, however, possible to evaluate these two differing worldviews with regard to one somewhat smaller metaphysical issue. This is the issue of reductionism.

Reductionism

The brief sketch of Aquinas's views given above makes clear that Aquinas's metaphysics is incompatible with reductionism, unlike SSP, which is committed to it. (For a defence of the claim that Aquinas's metaphysics rejects reductionism, see Chapter 1 of Stump (2003).) Although reductionism comes in many forms, they share a common attitude. In virtue of supposing that everything is reducible to the elementary particles composing it, reductionism holds that ultimately all macro-level things and events are a function only of things and events at the microstructural level. That is one reason why reductionism is often taken to imply a commitment to causal closure at the microphysical level.

For a helpful discussion of the general problem of reductionism relevant to the issues considered here, see Garfinkel (1993), who argues against reductionism by trying to show that reductive micro-explanations are often not sufficient to explain the macrophenomena they are intended to explain and reduce. He says,

> A macrostate, a higher level state of the organization of a thing, or a state of the social relations between one thing and another can have a particular realization which, in some sense, *is* that state in this case. But the explanation of the higher order state will not proceed via the micro-explanation of the microstate which it happens to *be*. Instead, the explanation will seek its own level . . .   (Garfinkel, 1993, p. 449)

Aquinas would agree, and Aquinas's account of the relation of matter and form in material objects helps explain Garfinkel's point. A biological system has a form as well as material components, so that the system is not identical to the components alone; and some of the properties of the system are a consequence of the form of the system as a whole. Garfinkel himself recognizes the aptness of the historical distinction between matter and form for his argument against reductionism. He says, 'the independence of levels of explanation . . . can be found in Aristotle's remark that in explanation it is the form and not the matter that counts' (p. 149). See also Kitcher (1993). Particularly helpful and interesting on this subject is a book by John Dupré (1995), who argues that causal determinism falls with the fall of reductionism.

One way to understand reductionism, then, is that it ignores or discounts the importance of levels of organization or form, as Aquinas would put it, and the causal efficacy of things in virtue of their form. This feature of reductionism also helps explain why it has come under special attack in philosophy of biology. (See, for example, Garfinkel (1993) and Kitcher (1993).) Biological function is frequently a feature of the way in which the microstructural components of a thing are organized, rather than of the intrinsic properties of the micro-components themselves. Proteins, for example, tend to be biologically active only when folded in certain ways, so that their function depends on their three-dimensional structure. But this is a feature of the organization of the protein molecule as a whole and cannot be reduced to properties of the elementary particles that make up the atoms of the molecule. (According to, for example, Richards (1991), for relatively small proteins, folding is a function of the properties and causal potentialities among the constituents of the protein, but 'some large proteins have recently been shown to need folding help from other proteins known as chaperonins'.)

One way to think about such recent anti-reductionist moves in philosophy is to see them as adopting a neo-Aristotelian metaphysics of a Thomistic sort. For Aquinas, a thing's configuration or organization, its form, is also among the constituents of things; and the function of a thing is consequent on its form.

On philosophical views such as these, a thing is not just the sum of its parts, reductionism fails, and there is no causal closure at the microphysical level. The component parts of a whole can sometimes explain *how* the whole does what it does. But *what* the whole does has to be explained as a function of the causal power had by the whole in virtue of the form or configuration of the whole.

An example drawn from neuroscience and psychology

Recent discoveries in neuroscience and developmental psychology suggest that we should go even further in this anti-reductionist direction. These discoveries suggest that in order to understand some cognitive capacities we need to consider a system that comes into existence only when two people are acting in concert, attuned to each other, as one.

Research on some of the deficits of autism have helped to illuminate such a system. Autism in all its degrees is marked by a severe impairment in what some psychologists and philosophers call 'mind reading' or 'social cognition'. We are now beginning to understand that mind reading or social cognition is foundational to an infant's ability to learn a language or to develop cognitive abilities in other areas as well.

For an infant to develop normally as regards mind reading, the infant's neural system has to be employed within the active functioning of a larger system composed of at least two persons, an infant and a primary caregiver. This system requires shared attention or joint attention between a child and its caregiver. Many lines of recent research are converging to suggest that autism is most fundamentally an impairment in the capacity for joint attention.

Trying to summarize his own understanding of the role that the lack of shared attention plays in the development of autism, distinguished psychologist Peter Hobson (2004, p. 183) says that autism arises 'because of a disruption in the system of child-in-relation-to-others'. By way of explanation, he says,

> My experience [as a researcher] of autism has convinced me that such a system [of child-in-relation-to-others] not only exists, but also takes charge of the intellectual growth of the infant. Central to mental development is a psychological system that is greater and more powerful than the sum of its parts. The parts are the caregiver and her infant; the system is what happens when they act and feel in concert. The combined operation of infant-in-relation-to-caregiver is a motive force in development, and it achieves wonderful things. When it does not exist, and the motive force is lacking, the whole mental development is terribly compromised. At the extreme, autism results.

On Hobson's views, then, autism cannot be explained apart from a complex system involving two human beings, an infant and its primary caregiver. Any attempt to explain this system in terms of reductionism and causal closure at the microphysical level would lose the understanding of the jointness in attention critical for normal infant development. On the contrary, as the phrase indicates, joint or shared attention cannot be understood even just by reference to one human being taken as a whole, to say nothing of the lowest-level components of a human being. Rather, it has to be understood in terms of a system comprising two human beings acting in concert. This system enables the shared attention which in turn enables the connection necessary for typical infant development.

The moral of the story

SSP supposes that all macro-phenomena are reducible to micro-level phenomena and that there is a complete causal story to be told at the micro-level. The converging lines of research in the sciences and several areas of philosophy, however, make a good case that reductionism is to be rejected. And if reductionism is rejected, then it is not true that the laws of higher-level sciences reduce to physics. It is not the case that everything is determined by the causal interactions at the level of the microphysical. And it is therefore also not the case that things at the macro-level are causally inert. Rather, causal power is associated with things at any level of organization in consequence of the configuration or form of those things.

In a metaphysical system of this anti-reductionist sort, the place of persons is not imperilled. In fact, even a human pair bonded in love, as a mother and child are, can be a sort of whole, with causal power vested in their bondedness.

If reductionism is rejected, as the new work in the sciences and in philosophy argues it should be, then with respect to this one issue the Thomistic worldview is more veridical and more worthy of acceptance than SSP is. By itself, of course, this conclusion certainly does not decide the issue as regards the central disagreement between SSP and the Thomistic view. It cannot adjudicate the issue regarding the ultimate foundation of reality. And so, as far as the evidence canvassed in this chapter is concerned, the central disagreement between SSP and the Thomistic view remains an open issue. Clearly, it is possible to reject reductionism and accept atheism.

For that matter, it is possible to reject atheism and accept reductionism. As I have described it, SSP is a secular view that combines contemporary scientific theories with certain metaphysical claims. But it is possible to have an analogue to SSP in which a reductionist scientific view of the world is combined with a commitment to religious belief, even religious belief of an orthodox Christian sort. That is, SSP can have a theistic analogue, which includes most of the scientific and metaphysical worldview of SSP but marries it to belief in an immaterial creator.

So, for example, consider Peter van Inwagen's explanation of God's providence. Trying to explain God's actions in the created world, van Inwagen says that God acts by issuing decrees about elementary particles and their causal powers: '[God's] action consists in His . . . issuing a decree of the form "Let that [particle] now exist and have such-and-such causal powers"' (van Inwagen, 1995, p. 49). For van Inwagen, apart from miracles, God's actions in the world consist just in creating and sustaining elementary particles and their causal powers. This, van Inwagen says, 'is the entire extent of God's causal relations with the created world' (van Inwagen, 1995, p. 44). On his view, miracles are a matter of God's supplying 'a few particles with causal powers different from their normal powers' (van Inwagen, 1995, p. 45).

For most people conversant with religious discourse in the Judaeo-Christian tradition, this religious analogue to SSP will seem a very odd mix. On their view, God not only issues decrees (about particles or anything else); God also cajoles, threatens, instructs, illumines, demands, comforts and asks questions. At the heart of all these activities is the direct interaction between persons of the sort Hobson was trying to explain. Even if, *per improbabile,* all this and more could be reduced to decrees about particles, the reduction would have lost the personal connection that in both Judaism and Christianity has been the most important element in the relations between God and human persons.

For these reasons, reductionism does not fit well with theism. I am not claiming that it is incompatible with theism. The point is only that there is something awkward or forced or otherwise implausible about reductionism in a theistic worldview. It isn't natural there, one might say. On a worldview that takes persons to be the ultimate foundation of reality, reductionism to the level of elementary particles is not really at home.

By the same token, it seems to me that the rejection of reductionism is harder to square with a worldview in which the ultimate foundation of reality is impersonal. Here too the issue is not the compatibility of the two positions. The point is rather this. The rejection of reductionism leaves room for the place ordinary intuition accords persons in the world. But, to me at any rate, the metaphysics that gives persons this place is more readily intelligible on a worldview that sees persons as the ultimate foundation of reality. Figuring out how to make it cohere with the picture Blackburn paints, even if we subtract reductionism from that picture, strikes me as much harder to do.

(This chapter is an abbreviated version of a much longer paper (Stump, 2015) which provides a consideration of more nuances and examples with regard to arguments about reductionism.)

4

The origin and end of the universe: A challenge for Christianity

DAVID WILKINSON

The joyous excitement of twentieth-century cosmology

A century ago on 18 November 1915, Albert Einstein made a discovery of which he wrote, 'For a few days I was beside myself with joyous excitement' (Einstein, 1998). For a number of years he had been engaged in quite tortuous work in trying to extend his work on the special theory of relativity to a more general theory. This involved deriving the field equations for gravity, which would describe how the geometry of space and time is shaped by the presence of matter and radiation. His joyous excitement came as he used these field equations to solve the decades-long puzzle of the small advance of the perihelion of Mercury (that is, the closest point of the planet to the Sun). Newton's gravitational laws needed something like another planet inside Mercury's orbit to explain this advance. Einstein applied his theory of gravitation and found that the advance was accounted for exactly without the need for other planets or other attempted 'fixes'.

While these field equations were still not in their correct final form, this incident was significant for a number of reasons. First, it illustrated the importance of astronomical observations to understanding the deepest structures of the universe. Second, in Einstein's joyous excitement was the sense that the universe is intelligible and that intelligibility is characterized by simplicity and beauty in the equations of physics. In fact the general theory of relativity was to be characterized by the Russian physicist Lev Landau as the 'most beautiful of theories'.

The consequences of the general theory were profound. It implied the existence of black holes, and predicted the gravitational bending of light and also gravitational waves. These waves have recently been discovered by the newly revamped Advanced LIGO (the Laser Inter-ferometer Gravitational-Wave Observatory), which surveyed 300,000 galaxies in the hope of catching the death spiral of two neutron stars.

Perhaps even more importantly, Einstein's field equations were at the centre of understanding the beginning and end of the universe. In 1917, Einstein applied his theory to the universe as a whole. At this stage it was assumed that the universe is static, neither expanding nor collapsing, and Einstein adapted his equations to this by including a cosmological constant. Yet in the period 1910 to 1930 astronomers such as Slipher and Hubble working on the redshift of galaxies began to show that the universe is expanding. Friedmann showed that this is consistent with Einstein's equations without the cosmological constant, and the Belgian priest and astronomer Georges Lemaître formulated the earliest version of the *Big Bang models*, in which our universe has evolved from an extremely hot and dense earlier state.

One of the great achievements of twentieth-century cosmology was this Big Bang model for the origin of the universe. It describes the expansion of the universe from a time when it was only 10^{-43} seconds old. At that stage, 13.8 billion years ago, the universe was an incredibly dense mass, so small that it could pass through the eye of a needle. This model is supported by the evidence of the redshift of galaxies, the microwave background radiation and the abundance of helium in the universe. Of course, like any scientific model it has some gaps. A large proportion of the universe is in the form of dark energy (over 70 per cent) and at the moment we have little idea as to what it is. Another 23 per cent of the universe is in the form of dark matter; we know it is there but we are not sure what it is. The fact that we know only a tiny fraction of what the universe is made of is somewhat embarrassing for cosmologists. Yet the power of science is that we know what we do not know, and we are able to design experiments at the Large Hadron Collider that might at least tell us what dark matter is.

Some questions are much more difficult. The standard model of the hot Big Bang describes the origin of the universe as an expansion from a singularity, that is, a point of infinite density. But that sin-gularity raises immediate problems. First, general relativity, which describes the expansion of the universe so well, suggests that time is not completely independent of space, and that gravity is then explained as a consequence of this space–time being curved by the distribution of mass-energy in it. Thus, the distribution of mass determines the geometry of space and the rate of flow of time. However, at a singularity there is infinite density and infinite curvature of space–time. General relativity is unable to cope with this infinity and predicts its own downfall; that is, the theory breaks down at the singularity. Second, general relativity as a theory is inconsistent with quantum theory. General relativity, which is extremely successful in describing the large-scale structure of the universe, needs to specify mass and its position in order then to describe the geometry and rate of flow of time. At a singularity, where the gravitational field is so strong and the whole universe is so small that it is on the atomic scale of quantum theory, it is believed that quantum effects should be important. Quantum theory, however, says that one can never know both the mass and the position of a particle without an intrinsic uncertainty. One cannot have both general relativity and quantum theory to describe a situation.

The *singularity problem* therefore is that general relativity is unable to give a description of the singularity; in other words, general relativity cannot explain the initial conditions for the expansion of the universe. Present scientific theories are thus unable to predict what will come out of the singularity. They can describe the subsequent expansion but are unable to reach back beyond an age of 10^{-43} seconds to zero. This limit of scientific theory, unable to reach back to the very beginning, is frustrating to physicists but attractive to some theologians. Is God needed to 'fix' the initial conditions of the universe? If science is unable to describe the initial moments, is this 'the gap' where God comes in to set the universe off?

Indeed, it is fair to say that the history of the Big Bang has always found the science entangled in some way with philosophy and theology. The *steady-state model* of the universe proposed by Bondi, Gold and Hoyle, attempted to avoid the 'beginning' implied by the Big Bang in part because they feared its theistic implications. Their attempt to argue against the evidence of the cosmological redshift and the microwave background was ultimately unsuccessful,

They were right, however, to fear that the Big Bang model would be baptized by certain religious thinkers. From the argument of the religious apologist – if the universe began with a Big Bang, then who lit the blue touchpaper of the Big Bang? – to more subtle attempts to revive the cosmological argument in temporal form, this model for the origin of the universe has been centre stage in the dialogue between science and religion.

Yet as the twenty-first century began to take over from the twentieth, significant thinking was taking place not just on the origin but also on the end of the universe.

A universe from nothing – but what do we mean by nothing?

Stephen Hawking's *The Grand Design* and Lawrence Krauss's *A Universe from Nothing* topped bestseller charts while making theological claims that God is not needed at the very first moment of the universe (Hawking and Mlodinow, 2010; Krauss, 2012). In fact, following the publication of *The Grand Design, The Times* (2 September 2010) led with the headline 'Hawking: God Did Not Create the Universe'.

However, against the media stereotypes claiming that these discoveries mean the death of a Creator, the interaction of Christian faith with the science of the origin and the end of the universe is much more complex and indeed fruitful. Sometimes these discoveries encourage a new dialogue with faith, and sometimes they lead to a new understanding of faith. The challenge of contemporary cosmology for Christianity is not a direct attack, but an opportunity to take science seriously in theological thinking and in building bridges between faith and culture.

In his earlier and worldwide bestseller, *A Brief History of Time*, Hawking had reacted against the feeling that, if our current scientific theories break down at 10^{-43} seconds after the initial event of the Big Bang, then this is the point where physicists should hand over to theologians. Indeed, in his most recent work he provocatively claims that 'Philosophy is dead. Philosophy has not kept up with modern developments in science, particularly physics' (Hawking and Mlodinow, 2010, p. 5). This reflects a widespread feeling among scientists that theologians and philosophers continue to assert generalizations about creation, rather than engaging with *specific* understanding of the theories that have attempted to extend the Big Bang model of the universe, such as inflation, string theory or M-theory. He is one of many scientists who rightly resist the trajectory of simply using God as an alternative to pursuing the scientific question.

Hawking attempts to use the laws of physics to explain not just the evolution of the universe but also its initial conditions. In order to do this one must bring quantum theory and general relativity together into a quantum theory of gravity. Such a theory, he suggests, can explain how the blue touchpaper of the Big Bang lights itself. The core of Hawking's theory, in John Barrow's phrase, is that 'once upon a time there was no time' (Barrow, 1993). According to Hawking, the universe does have a beginning but it does not need a cause, since in this theory the notion of time melts away. Hawking's universe emerges from a fluctuation in a quantum field. No cause as such is necessary.

Hawking believes that the best theory for explaining the universe's initial conditions is M-theory, which is in fact a whole family of different theories where each theory applies to phenomena within a certain range. It suggests 11 dimensions of space–time. However, for Hawking it also suggests that our universe is one in 10^{500} universes that arise naturally from physical law. And for him, 'their creation does not require the intervention of some supernatural being or god' (Hawking and Mlodinow, 2010, p. 8). It must be stressed that Hawking's thinking on this is not fully accepted by the rest of the scientific community. There are other proposals on how to deal with the problem of the laws breaking down, and it remains difficult to know whether quantum theory can be applied to the whole universe.

If Hawking's attempt to explain scientifically the first moment of the universe's history is indeed successful, then this does demolish a 'god of the gaps'. But the God of Christian theology is not a God who fills in any gaps of current scientific ignorance, nor one who interacts with the very first moment of the universe's history and then retires to a safe distance. Hawking's use of M-theory may eventually work, but the Christian theologian, while applauding enthusiastically, will also raise the question of where M-theory itself comes from. God is the one who creates and sustains the laws of physics, which science assumes but does not explain.

Such a god-of-the-gaps argument has been used in different contexts for centuries. However, it has a number of weaknesses. Augustine pointed out many years ago that the universe was created *with* time, not *in* time. Therefore to ask the question what came *before* the universe is an attempt to use the concept of time before time itself came into existence. In addition, the first-cause argument derives from a notion that the universe is a thing or event. It is easy to say that everything has a cause, but is the universe a thing or an event?

More importantly, as scientists explain more and more of the universe, there is a temptation to look for unexplained gaps in the knowledge of the natural world in order to find space for God. But this 'god of the gaps' is always in danger of becoming irrelevant as science fills in more of its own story. In contrast, the Bible understands that the whole universe is the result of God's working. God is as much at work at the first 10^{-43} second as at any other time. A scientific description of that moment in time does not invalidate its being the activity of God more than any other event. Indeed, the biblical images are not of a *deistic god* who breaks a bottle against the hull of the universe and then waves it off into the distance saying, 'Goodbye, see you on judgement day.' Paul in his letter to the Colossians, when speaking of Jesus, says, 'In him all things hold together.' This gives much more a picture of God as the one who keeps the universe afloat and together. God is the basis of the natural order, the basis of the physical laws. This is much more the God of Christian theism than the god of deism. Don Page, a long-time collaborator with Hawking, sums it up with these words:

> God creates and sustains the *entire* Universe rather than just the begin-
> ning. Whether or not the Universe has a beginning has no relevance
> to the question of its creation, just as whether an artist's line has a
> beginning and an end, or instead forms a circle with no end, has no
> relevance to the question of its being drawn.   (Page, 1998)

The end of the universe – but what about a new beginning?

If scientific work on the origin of the universe challenges Christian understandings of creation to reject deism and re-energize theism, work on the long-term future of the universe challenges a renewed emphasis on *new creation* as the central category of Christian hope.

This work was recognized recently in the award of the Nobel Prize for physics (Palmer, 2011). In 1998, astronomers began to look at distant supernovae explosions of stars. Their results showed something that was completely unexpected. The universe is accelerating in its rate of expansion due to some unknown type of force, the so-called 'dark energy' (Perlmutter et al., 1999, 2003; Riess et al., 1998). There had been no theoretical prediction of this, apart from Einstein's original inclusion of his cosmological constant in his solution of the equations of general relativity for the universe. It led to near panic among theorists, and to a range of possible explanations, none of which at the time of writing come anywhere near to a generally accepted understanding.

Yet the accelerating universe points to a future of futility for the physical and with it the end of the survival of intelligent life within the universe. An accelerated heat death is a bleak end. When the universe is 10^{12} years old, stars cease to form, as there is no hydrogen left. At this stage all massive stars have now turned into neutron stars and black holes. At 10^{14} years, small stars become white dwarfs. The universe becomes a cold and uninteresting place composed of dead stars and black holes.

Some physicists such as Freeman Dyson (1988) and Frank Tipler (1994) have tried to argue that the ability of humans in manipulating the environment will lead to the creation of forms of life able to survive such a universe. Dyson, for example, famously suggested that human intelligence could be downloaded into interstellar gas clouds, which could survive the low temperatures of a heat-death universe. However, while this may be possible in a universe slowing down in its expansion, it becomes increasingly impossible in an accelerating universe. Paul Davies is therefore correct in suggesting that an 'almost empty universe growing steadily more cold and dark for all eternity is profoundly depressing' (Davies, 2002, p. 48). It is a sobering thought that the optimism of science and technology in shaping our world for good is unable to find any hope in its own prediction of the futility of the end of the universe.

Some theologians will say that this is so far in the future that it is irrelevant, while others have concentrated their thinking on the future of the Earth, the individual believer or the Church. Yet here, Christianity can face the challenge and rediscover within its own tradition resources that give hope (Wilkinson, 2010). The theme of *new creation*, that is, a new heaven and Earth, is present within a range of biblical genres. This is not about some other-worldly exist-ence that has no connection with the physical universe. It is about God doing something with the totality of existence. At the same time, it is about something new, not about keeping this creation alive for as long as possible – which is the hope of such 'eschatological scientists' as Dyson and Tipler.

This new creation is a possibility because of a *Creator God*. The new creation is continually linked to God's original creative work, and hope for the future is built on an understanding of God as Creator. Whatever the circumstances, creation is not limited to its own inherent possibilities because the God of creation is still at work. The evidence of this work is focused in the resurrection of Jesus, which is also the model by which the continuity and discontinuity between creation and new creation are held together. If, as Paul argues, the resurrection is the first fruits of God's transformative work (1 Corinthians 15.20), then there should be both continuity and discontinuity in the relationship of creation and new creation, just as there was in the relationship of Jesus before the cross and Jesus risen. The empty tomb is a sign that God's purposes for the material world are that it should be transformed, not discarded. If resurrection affirms creation, then it also points forward to new creation. Continuity and discontinuity in the transformation of the physical universe may be located in the nature of matter, space and time. To take time as an example, the resurrected Jesus does not seem limited by space and time. In new creation, the continuity may be that time is real, but the discontinuity is that time no longer limits us in the way that it does in this creation. It could be argued that the resurrection body is characterized by decay's reversal, that is, a purposeful flourishing. In this creation, time is associated with decay and growth, but in new creation might time be simply about growth? We are therefore suggesting that our experience of time in the physical universe is a small and limited part of an ontologically real time that we might call *eternity*.

Such insights are offered as a structure for dialogue. They do not set out to map the biblical account exactly on to the scientific account, or to see them as completely independent. The Christian will come to the scientific description of the future of the physical universe with much to learn but also much to offer.

The distinguished cosmologist Martin Rees comments, 'What happens in far-future aeons may seem blazingly irrelevant to the practicalities of our lives. But I don't think the cosmic context is entirely irrelevant to the way we perceive our Earth and the fate of humans' (Rees, 2003, p. 4). This is a challenge to all theologians, not least those who take science seriously.

The joyous excitement of a God of creation and new creation

Albert Einstein is often quoted in discussions of science and religion, from 'Science without religion is lame, religion without science is blind' (Einstein, 1930) to 'God doesn't play dice with the world' (Hermanns, 1983, p. 58). Yet his own view of what he meant by God was not at all simple and does not map easily on to Christian understandings of God. He most probably came close to the pantheistic rather than personal God. He commented, 'My views are near those of Spinoza, admiration for the beauty of and belief in the logical simplicity of the order which we can grasp humbly and only imperfectly' (Hoffmann, 1972, p. 95).

This pantheistic God was also attractive to Sir Fred Hoyle, who moved from atheism to a belief that there was a deeper story to the universe, which culminated in his book *The Intelligent Universe*. One needs to respect such a view. It illustrates that science itself can pose questions but by itself cannot explain the role of God at the beginning and end of the universe. Or more importantly, it cannot describe the nature of God. After all, trying to use the Big Bang to demonstrate the existence of a creator may allow you to postulate a mathematically gifted designer but little more.

It is here that the relationship of science and theology becomes important (Barbour, 1997). The *conflict model* adopted by both New Atheists and six-day creationists in pitting science against religion seems to me to be oversimplistic in its view of both philosophy and actual history. Keeping science and religion independent of each other gives some insight into distinguishing between 'how' and 'why' questions, but again it is too simplistic. I am convinced that a *dialogue model*, although much more messy, is in fact more faithful to reality and much more fruitful. This recognizes that science and religion have different foci in exploring the universe, but there are areas of overlap where each can enrich the other and indeed pose questions for the other.

This dialogue is illustrated in the contemporary issues of the origin and end of the universe. For many Christian people, the Big Bang model gave an easy option to adopt a non-biblical 'god of the gaps' or deistic creator. But, since Hawking and others have raised the possibility of quantum gravity, Christians have been helpfully driven back to think through a doctrine of creation that sees God as the fundamental basis for the origin and continued existence of the laws of physics by which the universe evolves. This is not a god of the gaps, since science assumes the existence of those laws rather than explains where they come from. In this way, God is seen not as a creator who has long retired from active work, but as both *immanent and transcendent* in 'sustaining all things by his powerful word' (Hebrews 1.3, NIV). But in this there has to be a clear recognition that theology is bringing something to the table. The science by itself does not lead inevitably to this kind of picture. It only raises the question of where the laws of physics come from. Christian theology says that this understanding of the nature of God comes from a God who 'spoke to our ancestors through the prophets at many times and in various ways, but in these last days . . . has spoken to us by his Son' (Hebrews 1.1–2, NIV). It is in God's revelation in Jesus, both in historical events and as risen presence, that I find some insight into the nature of the Creator and creation.

It is in this revelation that I find hope in facing the cold desolate picture that science gives me for the future of the universe. Both as a scientist and a Christian, I want to take this seriously. A story such as a detective novel makes no sense if I just read the beginning, the middle or the end in isolation from the rest. Yet sometimes Christian theology has concentrated solely on the beginning of the universe, or on how God is at work in the world now. The end of the universe and the hope of new creation stemming from the resurrection of Jesus gives me a framework for life and indeed in viewing science. In the intelligibility and beauty of the origin of the universe I see science as a gift from God. Yet in the far future of the universe I see that science is not enough by itself to give me confidence and hope. While thanking God for science and rejoicing in it, I recognize that my faith has to be in the God of both creation and new creation.

This leads to a joyous excitement in scientific discovery which sometimes resonates with my belief in God and sometimes challenges my inadequate views of God. But it is also a joyous excitement to know that, whatever the mess of this world, the God that I see in the death and resurrection of Jesus is working towards a time when goodness, justice and love will triumph. That gives me confidence and energy to work for those things too and indeed to use science ultimately for those purposes.

5

Universe of wonder, universe of life

JENNIFER WISEMAN

New vistas

Marvelling at the heavens has always been an integral part of human existence, as far as we can know from history, art and ancient civilizations. But ever since Galileo recorded what he saw in the heavens with the aid of a telescope, human perceptions of the heavens and the Earth have been transformed for ever by our ability actually to investigate the cosmos. The ensuing Copernican Revolution changed our understanding of Earth as the focal point of the universe to a new paradigm of Earth as one component of a large, complex system of heavenly bodies.

Jumping ahead a few centuries, we are now witnessing 'new vistas' of the universe that are no less profound than that unveiled by Galileo. Perhaps the most radical in the past century is the realization, through theory and then observation, that the universe as a whole is changing, through the expansion of space itself. Proposed by the Belgian priest Georges Lemaître, and observed by the astronomer Edwin Hubble with his detection of galaxies all seeming to recede from us, the expansion of space points to the reality of a universe that is evolving from an earlier, extremely dense and hot state.

What exactly are *galaxies*? Edwin Hubble's observations showed that certain fuzzy entities seen with telescopes were so distant that they could not be situated within our own family of neighbourhood stars and gas clouds; these 'nebulae' must rather be separate entire galaxies outside our own Milky Way. We now know that there are hundreds of billions of galaxies filling the observable universe. They are collections of gas, dust and stars – more stars than our minds can fathom – held together by gravity. Galaxies like our own Milky Way each contain more than 200 billion stars (see Fig. 5.1). They also contain an even greater amount of unseen

Figure 5.1 Galaxy group Stephan's Quintet*
Credit: NASA, ESA and the Hubble SM4 ERO Team
Source: <http://hubblesite.org/newscenter/archive/releases/2009/25/image/c/format/large_web/>

dark matter, detected by its clear gravitational effects on visible stars and gas.

So, within just a few decades of recent human history, the scientific picture of the universe has moved from a simple, fairly static assembly of stars (including our Sun and solar system) in a single galaxy, to a mind-blowing collection of galaxies spread throughout an enormous, expanding universe that had an energetic start at some finite time in the past. The universe is immense, rich and dynamic. Where does life fit in? And, beyond science, is there some kind of purpose for the cosmos? Let us first delve just a bit deeper into what the galaxies, through telescopes, are telling us.

---

* This group of galaxies is located in the constellation Pegasus. While they appear close to one another, the galaxy in the upper left is actually much closer to us than are the others in the group. Note the varying shapes of the galaxies, the spiral arms in some, and the distorted shapes when two galaxies are merging together.

Galaxies, stars and planets: an active universe

Modern telescopes are revealing details of the heavens of which earlier generations would not even have dreamed. The galaxies that astronomers like Edwin Hubble observed decades ago come in many different shapes and sizes. Some are spherical, while others have a large fraction of their stars and gas entrained in majestic spiral arms anchored in a galactic core so dense with stars that the starlight can look like a blended beacon of white light.

What astronomers are now discovering is that galaxies harbour a lot of activity. Within spiral arms, for example, bright clouds of gas can emit colourful light due to the ionization of gas by young massive stars that have recently formed within the clouds. It turns out that star formation is an activity that has been ongoing since nearly the beginning of the universe. Simply put, a star is nothing more than a clump of mostly hydrogen gas in a galactic cloud that collapses into a very dense ball due to its own gravitational contraction. If there is enough mass, the pressure inside this collapsed gas clump will be high enough to enable the reaction known as fusion – hydrogen atoms combining to form helium, and releasing light. This is the birth of a star. Eventually, those photons of light escape from the surface of the newborn star, creating the starlight we can see.

Some galaxies appear to have mostly old stars, with very little gas left and thus very little continuing star formation. Other galaxies, often spirals, have a lot of gas, and star formation is vibrant. Entire galaxies can be drawn towards one another by their mutual gravitational pull, creating mergers that incite gas turbulence and a vibrant new burst of star formation.

We see star formation even in our own Milky Way galaxy, in regions relatively near (astronomically speaking) to our own solar neighbourhood. Taking advantage of new types of telescopes, astronomers have observed previously hidden stellar nurseries buried deep inside interstellar clouds by peering through the dust and gas with radio and infrared telescopes (see Fig. 5.2). In recent years, astronomers have detected not only infant 'protostars' in the process of forming, but also the presence of dense dark 'discs' of dusty material encircling and orbiting many of these stars around their equatorial zones. In some cases these flattened 'protostellar disc' zones are about

Figure 5.2 Star Cluster M13
Credit: NASA, ESA and the Hubble Heritage Team (STScI/AURA)
Source: <http://hubblesite.org/newscenter/archive/releases/star%20cluster/2008/40/image/d/format/large_web/>

the diameter of our solar system. The more mature discs are now recognized as the probable formation zones for planetary systems around stars, and protoplanetary disc studies are a hot topic in astronomy.

But do planets really exist around other stars? And how do planets form, with their rich composition of elements like carbon and iron, if stars are mostly made of hydrogen gas?

The answer to the second question comes from stars themselves. Stars are amazing little reactors, creating not only helium through the fusion of hydrogen in their cores, but also even heavier elements like carbon, oxygen and iron, through advanced reactions induced in their cores and during the death phases of the star. Not only is star formation an active process throughout the universe, but stellar death is also ongoing in every galaxy. Of course stars aren't actually 'alive', so by stellar death what is meant is the cessation of stable fusion reactions in the core of a mature star. After sometimes billions

Figure 5.3 The Crab Nebula*

Credit: NASA, ESA, J. Hester and A. Loll (Arizona State University)
Source: <http://hubblesite.org/newscenter/archive/releases/nebula/supernova%20remnant/2005/37/image/a/format/large_web/>

of years, most stars run low on hydrogen fuel in their cores, and eventually simply cool off after releasing their outer atmospheres during their final years of instability. Some stars more massive than our Sun become so unstable during their final days that they explode as spectacular supernovae with significant observable debris (see Fig. 5.3). In either case, old stars can release material into the interstellar environment that is enriched with elements heavier than hydrogen. Future generations of stars forming from this enriched interstellar gas will therefore also be enriched in their composition, and the disc zones forming around a young star can be fertile ground for the heavier elements and molecules that comprise planets, comets

---

* This nebula is a 6-light-year-wide expanding remnant of a star's supernova explosion that was first observed a thousand years ago. Debris from ageing or exploded stars enriches the interstellar environment and future generations of stars and planets with heavier elements.

and asteroids. Thus, stars themselves produce a variety of elements that enable and enrich future star and planetary systems, and even produce chemical elements needed for life.

In short, we wouldn't have planets and life as we know it without stars, and we wouldn't have stars without star formation, and we wouldn't have star formation without the gas in galaxies that is a product of the earliest phases of the universe itself. Cosmic activity is thus evident and important at every scale and stage. Next we shall examine whether planets are indeed a product of this active universe.

Planets everywhere: a fruitful universe

The actual existence of planets orbiting stars outside our own solar system has been nothing but speculation and imagination for all of human history – that is, until just the last three decades. Planets are very small compared to stars, and can be a billion times dimmer, lost in the glare of the parent star when observed by a distant telescope on or near Earth. Thus, it has taken some very clever technological advances to enable astronomers to detect only recently the presence of planets orbiting stars other than our Sun.

The earliest detections were accomplished not by observing these extra-solar planets, or *exoplanets*, directly, but rather by detecting the effects on their parent stars. As a planet orbits a star, the mutual gravitational interaction between planet and star causes the star to wobble slightly in its position; both planet and star are actually orbiting around a common 'centre of mass' located within the star but not at its centre. From the amount of stellar wobble observed with a special telescope detector, the planet's existence can be surmised, and its mass can be estimated. Hundreds of exoplanets were first detected this way.

Exoplanets that happen to be orbiting their parent star in a plane that is along our line of sight will pass in front of the parent star, eclipsing part of it from our point of view. These 'transiting' planets provide even more information, revealing the size of the exoplanet based on the amount of starlight the planet blocks during its transit. The search for transiting systems among 100,000 stars monitored by NASA's Kepler Telescope has produced detections of thousands of candidate planetary systems, and has confirmed the existence of planets ranging from Earth-sized and smaller to Jupiter-sized and

Figure 5.4 An artist's conception of the recently discovered exoplanet Kepler 452b

Credit: NASA Ames/JPL-Caltech/T. Pyle
Source: <www.nasa.gov/image-feature/soaking-up-the-rays-of-a-sun-like-star-artistic-concept>

larger. In fact we now have some amazing new statistics based on extrapolations from these observations: most stars in our galaxy have at least one planet, and most of these planets are smaller than Jupiter. A lot of them, perhaps most of them, are Earth-sized or slightly bigger, capable of having a solid surface rather than being a huge gas giant.

It is appropriate to pause for a moment and realize that we live in an amazing moment of human history. How grateful we should be for the technological advances that have enabled this incredible discovery that our galaxy is full of planets! (See Fig. 5.4.)

And now we are pushing the technology even further so as to learn more of the nature of these planets. Those that transit in front of their parent star enable us to measure something of their atmospheric composition, by measuring what frequencies of light from the star are absorbed by the foreground planet's atmosphere as the starlight passes through its outer regions on its way to our telescopes. So far, elements like sodium and nitrogen, and even molecules like water, have been detected this way, in a few exoplanet atmospheres.

Future telescopes will be able to discern whether there are planets with atmospheric compositions indicative of biological activity. The search for life beyond our solar system is a big curiosity driver of this quest to study exoplanets. In fact, an entire new field of science known as 'astrobiology' has sprung up with vigour to address interesting, inter-disciplinary questions about life in the universe: what are the environmental conditions needed for life? What kinds of environmental extremes can life endure on planet Earth? What kinds of environments on other planets could harbour and sustain life? What exactly *is* life? And how would we detect the presence of life remotely with telescopes?

If we see, for example, oxygen, methane and water vapour in an exoplanet atmosphere with our telescopes, it might indicate that some kind of life, perhaps primitive, is producing oxygen and methane and being sustained by water, such as is the case on Earth. The hope is that within the next few decades we will be able to survey hundreds of Earth-sized planets around nearby stars for signs of habitability or even of life.

The scientific attention right now on exoplanets and astrobiology is one focused mainly on the possibility of simple life, such as microbes, on exoplanets, as that is presumed to be the most prevalent if life beyond Earth exists. But of course human curiosity makes us wonder if advanced life fills the universe as well. Is there life of some kind everywhere?

This is not only a scientific question, but also a philosophical and theological one. What would it mean if life arose independently on other planets? Would this change humanity's view of its own significance? Would it conflict with, or support, theological ideas of creation? And if advanced life exists beyond Earth, would those life forms experience good and evil as we understand it? Would they embrace a belief in God? Christian belief would push these questions even further, given that the centre of this faith is the person of Jesus Christ, God of the universe in human form, who redeems needy and fallen humans on planet Earth. Is this same Saviour somehow redeeming advanced life in other forms on other worlds as well? These questions have been contemplated for centuries by religious leaders and philosophers. The Oxford scholar C. S. Lewis wrote about the possibility of life beyond Earth and its implications in his works of both theology and science fiction. More recently, the Vatican's Pontifical Academy of Sciences even hosted a major scientific conference on astrobiology, inviting leading scientists and scholars to discuss the latest advances in the scientific search for life beyond Earth, and its implications, with an air of excitement. Father José

Funes, Director of the Vatican Observatory, is quoted by Riazat Butt in *The Guardian* reflecting:

> Why is the Vatican involved in astrobiology? ... although astrobiology is an emerging field and still a developing subject, the questions of life's origins and of whether life exists elsewhere in the universe are very interesting and deserve serious consideration. These questions offer many philosophical and theological implications.   (Butt, 2009)

The big picture: an evolving universe

Let us now 'zoom out' from the local view of life on individual planets back to the 'macro' view of the entire universe. We have now seen that recent technological advances in telescopy have enabled astronomers within the last century to discern that the entire universe is expanding from a beginning era in time, and more recently that exoplanets really do exist in abundance, accompanying many (and probably most) stars in our own galaxy, and presumably in others as well.

What is emerging from these new advances is perhaps the most profound revelation of astronomy: our entire universe has been changing, evolving and maturing for about 14 billion years, transforming from energy and the first simple matter into complex galaxies and star systems, with at least one environment (Earth!) where diverse life thrives and advanced life contemplates its own existence and purpose.

There are several independent lines of evidence that our universe had a 'beginning', and it is amazing that they all point to the same age for the universe. The first crude estimate of this age can be found by taking measurements of the current expansion rate of the universe and simply 'rewinding' the expansion in a thought experiment to calculate how long this could have lasted. It's not quite so straightforward: recent studies indicate that the expansion of the universe is actually accelerating due to the outward push of some force or energy termed for now *dark energy*, for lack of a better understanding of what it actually is. Yet for the first few billion years, following a brief but profound period of inflation, it appears that the expansion of the universe was decelerating. Taking all this into account, the total time of the universe's expansion would be nearly 14 billion years, in agreement with estimates derived independently by looking at the character of the *cosmic microwave background* radiation. This is the incredible remnant of radiation filling the universe, left over from the 'Big Bang' burst of energy from which the universe has evolved. By the way, both phenomena – the cosmic microwave background radiation, and the acceleration of the expansion of the universe – earned their discoverers Nobel prizes. The third line of evidence comes from looking at stars and deducing their ages through astrophysical models involving their compositions, temperatures and radiation. The oldest stars turn out to be about 13 billion years old, slightly younger than the age of the universe as a whole, as would be expected. Therefore, while early in the twentieth century there was still an active debate among astronomers as to whether the universe had a 'beginning' or had been existing for ever in a 'steady state', there is now nearly unanimous agreement that our universe had an energetic beginning about 14 billion years ago, or 13.8 to be even more precise.

As an aside, according to some multiverse theories, other unobservable universes outside our own may also exist and have independent beginnings and different internal physical laws, and the whole ensemble of universes might not have something we would term a shared 'beginning'. While this construct may or may not exist in reality, it need not be a theological problem: the God of Judaeo-Christian belief is outside 'time', and would be responsible for anything that exists, including a multiverse. Time as we experience it would have been a product of the Big Bang in our own universe.

Our own universe, however, seems to have evolved in spectacular fashion, and, remarkably, we can actually observe its evolution thanks to the phenomenon of *light travel time*. Everything in the universe we observe is appearing to us the way it did when the light first started its journey, not as the source actually is at the present moment. For nearby sightings in our daily lives, the light travel time is inconsequential. But for observing other star systems and especially other galaxies, this plays a huge role in interpreting what we see (we must also account for reddening of light due to its journey across stretching space). Our nearest neighbour grand spiral galaxy, Andromeda, is about 2 million light years away. We are today seeing Andromeda as it was 2 million years ago.

With our best telescopes, we can now see extremely faint galaxies that are shining to us from billions of light years away. In fact we can see some galaxies shining to us from their infancy within the first '0.8' of the '13.8'-billion-year history of the universe. We can examine both the morphology and the composition of galaxies at different distances from us, thereby sampling different time epochs of the universe. What we find is astounding: the most distant (and therefore most infant) galaxies are small, and are comprised of stars and gas made up mostly of hydrogen and a little helium, with not much else. Most galaxies closer to our own Milky Way in space and time are larger, due to mergers of smaller galaxies, and have star-system compositions that include a significant amount of heavier elements, like carbon, iron and oxygen. Generations of stars have come and gone in these more mature galaxies, and these 'stellar factories' have produced the heavier elements that enable subsequent star systems to form with dust and planets and even potentially to sustain life.

A purposeful universe

The main theme of this book is to investigate whether science and faith need one another. So far, this chapter has examined incredible scientific discoveries about the universe, with only a few mentions of related philosophical and theological interests. But is there a role for faith as we contemplate this evolving universe? Perhaps faith plays the important role that science alone cannot play in addressing the big philosophical questions of 'Why?' and 'Is there a purpose to the universe?' and 'Is there a purpose for our lives in the universe?' The sciences can better inform these types of question, and faith can better answer them. As we have seen, it is now becoming apparent that the entire universe has been evolving in a direction that has enabled the conditions needed for life, even advanced life, to exist and to thrive, at least on one planet. And in fact there may be life-bearing planets filling the universe. We humans are here – we are contemplating our own existence, and, as we do so, our bodies contain elements that were literally forged in previous generations of stars.

Thus, modern astronomy is revealing that *evolution* is not just a feature of biological systems but is actually a profound frame for the entire history of the universe. One could reasonably conclude, through eyes of faith informed by this science, that the evolution of the universe from the Big Bang, through the early development of gas, stars and galaxies, through generations of stars and the production of heavy elements and complex molecules and planets, through the appearance and messy evolution of life, to complex ecosystems that include contemplative advanced life, implies a purpose of cosmic proportions.

Such a conclusion is not without challenges. Some scientific projections of the distant future envisage a cold, dark universe, with galaxies becoming increasingly isolated from one another, dominated by black holes and darkness. Here perhaps it is worth positing that those same eyes of faith mentioned above could trust in a larger (though yet unseen) purpose for the universe yet to come, based on the grand evolutionary history already attained. As a profound basis of such faith, the writer of the Gospel of John proclaims that 'all things were made through God', who is described as the *logos*, or the 'Word'. 'In him was life . . . The Word became flesh and dwelt among us. We have seen his glory . . .' So, perhaps *purpose* can be best seen in terms of *relationship*, as this penetrating passage implies.

Let us close with a simple recognition also of *cosmic beauty*. The evolution of the universe according to discernible laws of physics portrays a kind of elegant beauty, yet with variety – no two galaxies look exactly alike, and yet they are all beautiful and astounding to our eyes and minds. We recognize the reality of ugliness and injustice and evil on our planet, in part because they are in such contrast to the beauty of love, and the beauty we see in the cosmos. Perhaps the discovery of life flourishing across the universe will some day show us another aspect of cosmic beauty and wonder. So, let us continue exploring with curiosity, diligence and awe.

6

Evolution, faith and science

KENNETH R. MILLER

Voices against evolution and faith

In 2005, the first amendment to the United States Constitution was put to the test in a federal courtroom. The issue was not freedom of speech, freedom of the press or the right to assemble, all of which are guaranteed by that amendment. Rather it was the amendment's language forbidding Congress, and by extension other governing bodies, from passing any law 'respecting an establishment of religion'. Eleven parents in the small town of Dover, Pennsylvania, had gone to court alleging that the town's school board had violated that provision by requiring that students be taught about 'intelligent design' (ID), a religiously inspired alternative to the theory of evolution.

I took the stand as an expert witness during the first two days of that seven-week trial, testifying as to the scientific standing of the theory of evolution as well as to the intellectual bankruptcy of the flawed arguments made on behalf of 'intelligent design'. The outcome of that trial, featured in four books (see, for example, Lebo, 2008) and a 2006 BBC programme entitled *A War on Science*, was a resounding victory for the parents and the scientific community that had supported them. It was also a signal defeat for the anti-evolution movement in the USA.

Unfortunately, the attention received by the Dover trial lent itself to the temptation of oversimplification. Like similar confrontations, including the famous 1925 Scopes 'monkey trial', it was far too easy to characterize the proceedings as a 'God vs Science' confrontation. Critics of the decision were quick to see it as a blow against religious free expression, and an example of the willingness of Darwinist elites to censor competing ideas.

In reality, it was none of these things, but the perception of scientific hostility to religion in general and Christianity in particular nonetheless lies at the very root of anti-evolution movements in the United States and Europe. It is certainly true that these movements have had plenty of help in making the point that evolution can be used as a weapon against religious faith. An oft-cited book review written by the philosopher David Hull is one of their prime examples:

> Whatever the God implied by evolutionary theory and the data of natural history may be like, He is not the Protestant God of waste not, want not. He is also not a living God who cares about His productions. He is not even the awful God portrayed in the book of Job. The God of the Galápagos is careless, wasteful, indifferent, almost diabolical. He is certainly not the sort of God to whom anyone would be inclined to pray. (Hull, 1991)

Hull's point seems to be that God, if he exists, is a pretty nasty fellow. His evidence, naturally enough, is nature red in tooth and claw, a natural world containing an evolutionary process he regards as 'rife with happenstance, contingency, incredible waste, death, pain, and horror' (Hull, 1991). No God could allow such horrors, so there can be no God, according to Hull and his view of evolution. The distinguished evolutionary biologist Richard Dawkins has been even more explicit on this point, making it clear that his view of the evolutionary process is at least as bleak as Hull's:

> In a universe of blind physical forces and genetic replication, some people are going to get hurt, other people are going to get lucky, and you won't find any rhyme or reason in it, or any justice. The universe we observe has precisely the properties we should expect if there is, at bottom, no design, no purpose, no evil and no good, nothing but blind pitiless indifference. (Dawkins, 1995, p. 133)

Wonder and the purpose of existence

Curiously, lost in this rush to assert the pointlessness of life is the sense of wonder with which Darwin himself approached the evolutionary process, namely, that 'from so simple a beginning endless forms most beautiful and most wonderful have been, and are being, evolved' (Darwin, 1859, pp. 459–60). The modern view, it would seem, has wrung the sheer delight out of Darwin's vision, and enlisted it in a philosophical assault against religion. Reading such pronouncements, one cannot help but notice how neatly they fit into the strategic plans of the anti-evolution movement. Indeed, it is by making evolutionary science the enemy of God, according to the University of California emeritus professor Phillip Johnson, that the religious aims of the ID movement can be realized. Writing in *Church and State* magazine, Rob Boston summarized Johnson's views this way:

> The objective [according to Johnson] is to convince people that Darwinism is inherently atheistic, thus shifting the debate from creationism vs. evolution to the existence of God vs. the non-existence of God. From there people are introduced to 'the truth' of the Bible and then 'the question of sin' and finally 'introduced to Jesus'.
>
> (Boston, 1999)

Considerations such as these led the columnist Madeleine Bunting of *The Guardian* to explain to her readers that 'the intelligent design lobby thanks God for Richard Dawkins', noting that 'Anti-religious Darwinists are promulgating a false dichotomy between faith and science that gives succour to creationists' (Bunting, 2006). Exactly so.

As an experimental scientist, what I find especially noteworthy in pronouncements from individuals such as Hull and Dawkins is an assumption implicit in their use of evolutionary science in philosophy. That assumption is that science alone can lead us to truth regarding the purpose of existence – which is, of course, that it does not have one. This may or may not be true, of course, but it is not a scientific statement because it is not testable by the methods of science. Indeed, David Hull's pronouncements about the 'waste' and 'horror' of existence have no more scientific standing than a faith-based assertion one might make echoing the words of Darwin to describe the profusion of 'endless forms most beautiful and most wonderful' in the world of life.

In fact, nearly all biologists would agree that the capacity for evolution, as well as life itself, is built into the fabric of the natural world as an inherent part of the physics and chemistry of matter. If this is true, then the apparent chaos of that world actually contains the seeds to produce, by its own means, the order, design and beauty of life in which we so delight. As a result, the Christian notion that we live in a universe of meaning and purpose is validated rather than contradicted by the ever-expanding evolutionary possibilities of existence. John Haught, a Georgetown University theologian, has written extensively on this point (see, for example, Haught, 2001).

The nature of evolution to a person of faith

But doesn't evolution contradict the role of God as Creator, described so dramatically in the book of Genesis? An answer to that question turns, of course, on what we suppose the Creator might have fashioned. If we misconstrue Genesis as natural history, we find profound contradictions between its creation story and the modern sciences of astronomy, geology and biology. Heaven and Earth were not fashioned 6,000 years ago, all living organisms did not appear simultaneously, and the Earth's geological formations were not laid down in a single worldwide flood. These claims were refuted well before Charles Darwin's time, and the realization that life on Earth is not only ancient, but also ever changing, forms the foundation of the geological sciences.

What the modern theory of evolution does show is that the origins of all species, including our own, are found in natural processes that can be observed and studied scientifically. In other words, evolution demonstrates that our own existence is woven into the very fabric of the natural world. Seen in this light, the human presence is not a mistake of nature or a random accident, but a direct consequence of the characteristics of the universe. What evolution tells us is that we are part of the grand, dynamic and ever-changing fabric of life that covers our planet. To a person of faith, an understanding of the evolutionary process only deepens our appreciation of the scope and wisdom of the Creator's work.

The particular solution for people of faith, therefore, is not to oppose science, but to develop an understanding of science that is in harmony with religious faith. Taking up this task, I am convinced, is the key to making peace between science and religion, a peace that is much to be desired. I am hardly the first person to make this point. The notion that religion must respect the finding of scientific reason is, in fact, a traditional Western view that has been expressed by many writers in the Christian tradition, none more eloquently than St Augustine:

> Even a non-Christian knows something about the Earth, the heavens . . . the kinds of animals, shrubs, stones, and so forth, and this knowledge he holds to as being certain from reason and experience. Now it is a disgraceful and dangerous thing for an infidel to hear a Christian, presumably giving the meaning of Holy Scripture, talking nonsense

on these topics; and we should take all means to prevent such an embarrassing situation, in which people show up vast ignorance in a Christian and laugh it to scorn.

(Augustine, 1982, Book 1, Chapter 19, p. 16)

This remarkable passage points out that believers and non-believers alike have equal access to observations of the natural world. Therefore, nothing could be worse for people of faith than to defer to the Bible as a source of scientific knowledge that contradicted direct, empirical studies of nature. Augustine, one of the most prolific and influential of the early Christians writers, got the relationship between Scripture and empirical science exactly right. He warned of the danger inherent in using the Bible as a book of geology, astronomy or biology, admonishing the faithful that to do so would hold the book up to ridicule and disproof. To Augustine, the eternal spiritual truth of the Bible would only be weakened by pretending that it was also a book of science.

For Christians today, the scientific successes of evolutionary theory present a genuine opportunity to come to grips with the reality of the natural world that gave rise to us. That science, no question about it, presents genuine challenges to religion, but it also provides religion with an extraordinary opportunity to inform and enlighten the scientific vision of our existence. As if to illustrate a pathway to such understanding, several years ago one of my scientific friends sent me this passage, and asked me to guess its author:

> According to the widely accepted scientific account, the universe erupted 15 billion years ago in an explosion called the 'Big Bang' and has been expanding and cooling ever since ... In our own solar system and on Earth (formed about 4.5 billion years ago), the conditions have been favourable to the emergence of life. While there is little consensus among scientists about how the origin of this first microscopic life is to be explained, there is general agreement among them that the first organism dwelt on this planet about 3.5–4 billion years ago.
>
> (International Theological Commission, 2004, paragraph 63)

The 'author' of that brief but straightforward account of scientific natural history was, according to my colleague, Pope Benedict XVI. To be perfectly accurate, he wasn't exactly the 'author', since the passage actually comes from the 2004 report of a committee known as the International Theological Commission, but Joseph Cardinal

Ratzinger (later to be Pope Benedict) did indeed supervise the work of the Commission, and clearly approved its final form. Significantly, the report goes on to make specific comments about evolution that relate to the evolution–creation struggle:

> Many neo-Darwinian scientists, as well as some of their critics, have concluded that, if evolution is a radically contingent materialistic process driven by natural selection and random genetic variation, then there can be no place in it for divine providential causality ... But it is important to note that, according to the Catholic understanding of divine causality, true contingency in the created order is not incompatible with a purposeful divine providence. Divine causality and created causality radically differ in kind and not only in degree. Thus, even the outcome of a truly contingent natural process can nonetheless fall within God's providential plan for creation.
>
> (International Theological Commission, 2004, paragraph 69)

Evolution is indeed a 'truly contingent natural process', and the Commission's clear statement that such a process can fall within the sphere of divine causality is nothing more than a reaffirmation of the teachings of Aquinas and other Christian writers on divine and natural causality. This kind of clarity, unfortunately, is remarkably rare in public statements on both sides of the religion and science debate today.

Science and religious faith

All too often, people of faith have sought to use scientific findings to justify and defend specific interpretations of Scripture and religious doctrine. This has been a principal tactic of the creationist movement, which has attempted to twist scientific findings to support pseudoscientific readings of the book of Genesis. In taking this point of view, they look to science to justify faith, indirectly and unintentionally admitting that religious beliefs must turn on the latest findings from the field or laboratory. This approach places faith in a subordinate position to science, and helps to explain why 'scientific' creationists express hostility to ideas such as cosmic expansion and evolution. It is, I believe, a profound misrepresentation of the true relationship between faith and science. Consider, for example, the embrace of scientific reason expressed by Theodosius Dobzhansky, a Christian and one of the foremost evolutionary geneticists of the twentieth century:

The organic diversity [of life] becomes, however, reasonable and under-
standable if the Creator has created the living world not by caprice but
by evolution propelled by natural selection. It is wrong to hold creation
and evolution as mutually exclusive alternatives. I am a creationist and
an evolutionist. Evolution is God's, or Nature's method of creation.
Creation is not an event that happened in 4004 BC; it is a process that
began some 10 billion years ago and is still under way.

(Dobzhansky, 1973)

Rather than looking to science to confirm a primitive understanding
of life and cosmos expressed thousands of years ago in a pre-scientific
age, Dobzhansky understood science as a way to refine and expand
our understanding of the Creator's power and majesty. This, I would
suggest, is a model for the proper relationship between science and
faith. A similar understanding was expressed more recently by Guy
Consolmagno, an astronomer and Jesuit brother, appointed head
of the Vatican Observatory. Interviewed by *Astrobiology* magazine,
Consolmagno stated:

The trouble is that some people think they can use science to prove
God. And that puts science ahead of God; that makes science more
powerful than God. That's bad theology. In fact, some philosophers
have said that's what led to atheism in the eighteenth century – the
fallacy of the God of the gaps. You say, 'I have no idea how this could
have happened. It must have been God's design'. And then fifty years
later, somebody explains how it did happen, and you say, 'I don't need
God anymore'. If your faith is based on science, that's a very shaky kind
of faith. My belief in God is not because of something I've seen in
science. But I can turn it the other way around and say, 'I believe
in science because of my faith in God'.   (Consolmagno, 2005)

The historical roots of modern science lie not in a rejection of faith,
but rather in the conviction that exploration of the natural world
is an act of praise and worship. As Aquinas and other Christian
philosophers have emphasized, faith and reason are both gifts from
God, and as such they should be complementary. In many ways,
I would argue that science itself, regardless of the religious beliefs
of its practitioners, is based on two great elements of faith. The first
is that a genuine universe exists and can be understood by rational
scientific inquiry. The second is that knowledge of that universe,
gained through science, is to be preferred to ignorance. Albert Einstein,
although not a theist, echoed these sentiments when he wrote:

While it is true that scientific results are entirely independent from religious or moral considerations, those individuals to whom we owe the great creative achievements of science were all of them imbued with the truly religious conviction that this universe of ours is something perfect and susceptible to the rational striving for knowledge.

(Einstein, 1954, p. 52)

Ultimately, the religion and science debate continues because of a deep antagonism between extremists on both sides of the issue. The solution is not to split the difference, but to come to a genuine understanding and appreciation of the true depth of scientific and religious thought on the issues at hand. In the specific case of evolution, the sophistication of Christian thinking on natural processes and the divine will is routinely underestimated by those who would use science as a weapon against faith. Conversely, the Christian community often fails to appreciate the self-critical nature of science and the clear recognition of most scientists as to the limitations of scientific inquiry. In the final analysis, both sides may come to realize, as Charles Darwin did, that there is indeed beauty, wonder and even grandeur in the evolutionary view of life.

(Portions of this chapter have been excerpted from Kenneth Miller's previous essays, including a 2009 article in the March/April issue of *BBC Knowledge*, and his 2009 Terry Lecture at Yale University, published in *The Religion and Science Debate: Why Does It Continue?* (Yale University Press, 2009).)

7

Evolution and evil

MICHAEL J. MURRAY AND JEFF SCHLOSS

Setting the scene: the problem of evil

The 'problem of evil' has been recognized for millennia as a serious challenge to faith in a good God – by both those who accept and those who reject theism. However, over the last 150 years or so the issue is purported to have taken on even graver significance in the light of the natural sciences. Emerging discoveries about the living world in general and evolution in particular seem to have cast nature not merely as including evil, but as being essentially characterized by it – perhaps even to the utter exclusion of natural goodness. The iconic representation of this view is often ascribed to Alfred Tennyson's epic poem, *In Memoriam A. H. H.*:

> This round of green, this orb of flame,
> Fantastic beauty such as lurks
> In some wild poet when he works
> Without a conscience or an aim . . .
>
> Who trusted God was love indeed,
> And love Creation's final law,
> Tho' Nature red in tooth and claw
> With ravine shriek'd against his creed . . .
>
> (Tennyson, 1849)

Although Tennyson wrote before Darwin, he was informed and troubled by new findings in the sciences that seemed to subvert nature's testimony of beneficence. These findings and their posited entailments – adopted and disseminated by many of the leading biologists working around that time – both fuelled and reflected a backlash against a Romantic view of the natural world. For example, Charles Darwin famously wrote, 'What a book a devil's chaplain might write on the clumsy, wasteful, blundering, low, and horribly cruel works of nature' (Darwin, 1856). And in a prominent essay on the struggle for existence, T. H. Huxley ('Darwin's bulldog') claimed, 'The animal world is on about the same level as a gladiators' show . . . The Hobbesian war of each against all is the normal state of existence' (Huxley, 1888). In an earlier and less well-known reflection, Huxley laments:

> You see a meadow rich in flower and foliage, and your memory rests upon it as an image of peaceful beauty. It's a delusion . . . Not a bird twitters but is either slayer or slain . . . not a moment passes in that a holocaust, in every hedge & every copse battle, murder, & sudden death, are the order of the day.   (Huxley, 1887)

The twofold claim is that:

1 nature doesn't just contain suffering but is in all-out war;
2 this 'underground warfare', unseen by pietistic sentimentalism, has been exposed by discoveries made possible by scientific rigour.

Similarly, prior to publishing *The Origin of Species* Darwin maintained, 'De Candolle, in an eloquent passage, has declared that all nature is at war . . . Seeing the contented face of nature, this may at first be well doubted; but reflection will inevitably prove it to be true' (Darwin, 1858).

In this chapter we wish to assess, first, whether this characterization of nature is a necessary conclusion from *evolutionary biology* and, second, to the extent that widespread conflict and suffering are an undeniable feature of nature, what resources exist for *theodicy*.

Those who argue that evolution increases the force of the problem of evil typically focus on one or more of three considerations. First, evolutionary history greatly expands the *extensiveness* of natural evil. Second, evolutionary process re-characterizes the *role* of natural evil and makes it a central element of how nature works, rather than just a by-product or post hoc feature. Third, in addition to expanding the extent and accentuating the role of evil, in the view of many, Darwinian evolution precludes the very existence of either natural goodness or progress: Tennyson's 'without a conscience or an aim'. Without dismissing the testimonial ambiguity of the natural world or diminishing the serious challenge of natural evil itself – recognized from the early Hebrew Scriptures up through virtually all contemporary philosophers of religion – we want to ask:

1 Are these claims in fact new?
2 What are the necessary implications of a specifically evolutionary understanding of the natural world?

Implications of evolutionary biology

The extent of natural evil: biological armament and mass extinction

Two discoveries of nineteenth-century science especially amplified concern about the extensiveness of evil: biological armaments in species interactions and patterns of massive extinction in the fossil record. One example of the first is the discovery of *parasitoids*, which are parasites whose life history necessarily involves killing their host, rather than co-adapting to live with the host. The iconic example of this is the ichneumon wasp, which lays eggs in a caterpillar. Those eggs then hatch into larvae that eat the flesh of the caterpillar before emerging, thus destroying their living host from the inside out. Reflecting on this, Darwin concluded, 'I cannot persuade myself that an omnificent and omnipotent God would have designed and created the Ichneumonidae with the expressed intention of their feeding within the living bodies of the caterpillars' (Darwin, 1860). The ichneumon discovery went viral (or 'went parasitoid') in nineteenth-century popular discussion and – in spite of the fact that nobody thought the host actually suffered – it became emblematic of the post-Romantic cold shower in the facts of nature.

But this is just one example of manifold, fascinating discoveries of biological 'warfare', many of which emerged out of encounter with tropical biota. Often referred to as an escalating biological 'arms race', strategies include slave making, infanticide and matricide, chemical warfare against competitors or predators, false advertising, competitive sabotage, concealed weapons, ambush attacks, suicide defence and so on.

Describing the scale of extinction in the fossil record also expanded our sense of the extensiveness of natural evil. In fact, when fossils were first discovered they precipitated a debate over whether they were actually the remains of living organisms, or whether they were residues of inanimate physical processes, perhaps similar to crystal formation. The latter conclusion seemed preferable to one that entails so much death and destruction in life's history. But, once consensus was reached that fossils are remains of organisms, it raised powerful concerns. As Tennyson lamented, one could no longer take solace that above the level of individual suffering, God was somehow still concerned at the level of the species, or type:

> Are God and Nature then at strife,
> That Nature lends such evil dreams?
> So careful of the type she seems,
> So careless of the single life ...
>
> 'So careful of the type?' but no.
> From scarped cliff and quarried stone
> She cries, 'A thousand types are gone:
> I care for nothing, all shall go ...'
>
> (Tennyson, 1849)

Reacting to the kinds of discoveries described above, the contemporary philosopher David Hull remarks,

> The evolutionary process is rife with happenstance, contingency, incredible waste, death, pain, and horror. Whatever the God implied by evolutionary theory and the data of natural history may be like, He is not ... a loving God who cares about His productions ... [He is] careless, wasteful, indifferent, almost diabolical. He is certainly not the sort of God to whom anyone would be inclined to pray.
>
> (Hull, 1991)

Of course the challenge of natural evil should not be avoided. But neither should it be exaggerated or wrongly attributed to a *scientific* insight. The latter caution merits three comments. First, none of the above observations are entailments of or made worse by evolutionary theory. They are brute observations of the natural world, which were recognized before Darwin and which would remain true – and remain troublesome – even if God created each species individually. Second, it is not at all clear how 'extinction' magnifies the problem of natural evil. The sum total of death, with or without extinction, is precisely one per living creature. And to claim that it is 'wasteful and careless' for a species not to be immortal but rather to take turns in the developmental sequence of life is a conclusion in search of an argument. It assumes – without rationale – that it is a 'waste' for anything to exist if it does not exist for ever, however majestic the serial cascade of life may be. Third, while our understanding of competitive interactions between and within species has indeed expanded, so too has our understanding of *cooperative interdependence*. Interestingly, though many proposals concerning the importance of interdependence were around at the time of Darwin (for example, the existence of co-operative symbiosis), they were initially resisted as being incompatible with the fundamentally competitive struggle of life.·

The role of natural evil: competition and cooperation

However, maybe it is not the sheer *amount* of evil, or even the *ratio* of evil to goodness, but rather the *role* of natural evil that is most significant. That is the second issue, which appears to be a more serious problem that really does derive from how the mechanism of evolution is widely understood. Natural selection is often represented as involving stringent competition or an unrelenting 'struggle for existence'. This competitive struggle is not an unfortunate intrusion into the workings of nature, but is the very wellspring of diversity and engine of evolutionary change. And it involves not only suffering and death, but a merciless 'survival of the fittest': Tennyson's 'I care for nothing, all shall go' (Tennyson, 1849).

Again, competition and death are genuinely vexing – and also long-recognized – aspects of the living world. But it is not so simply the case that they are the engines of diversity or the drivers of evolution. First, although often confused, competition and natural selection are actually not the same thing, nor are they even necessarily conjoined. 'Competition' is strictly defined as two individuals negatively influencing one another's fitness. 'Natural selection' is simply differential reproduction. You can have competition without natural selection, and more importantly, natural selection can operate without competition. (Think of a novel trait that increases fitness by allowing exploitation of a new resource, without at all decreasing the fitness of those without the trait.) Second, much evolutionary change occurs under conditions of relaxed selection, such as the opening up of new habitats. And third, even though competitive struggle is undeniably a driver of evolution under certain circumstances, so too is *cooperative synergy*, which can facilitate the emergence of novel adaptations and new levels of function. In a recent scientific review, the Harvard evolutionary biologist Martin Nowak concludes, 'Cooperation can now be seen as an additional, fundamental force in evolution' (Ohlson, 2012). And although we don't want to fall into the sentimental oversimplification of 'love Creation's final law', the other polarity of 'warfare Creation's ultimate end' is equally untrue. In popular writing, Nowak speaks of the 'snuggle for existence' operating along with the struggle for existence.

The existence of progress and natural goodness

But what reason is there to believe this is really true? Several decades ago, critics claimed that evolution not only involves a great deal of evil and death, but is ultimately attended by an utter dearth of goodness. This is the third issue, one highlighted by the Tennyson phrase, 'without a conscience or an aim'. The purported twofold problem involves an absence of 'aim' or direction in the unfolding of life, and the exclusion of 'goodness' or natural beneficence. Regarding the first, the evolutionary biologist Stephen Gould asserted, 'Progress is a noxious, culturally embedded, un-testable, nonoperational, impractical idea that must be replaced if we wish to understand evolutionary history' (Gould, 1988, p. 319). In fact, he not only rejected the notion of progress, but made the stronger claim that evolution lacks significant directional trends altogether, characterizing it as a man staggering out of a bar, dead drunk, and stumbling into the gutter. And where he ends up (i.e. the second issue) is the wrong side of town. Regarding natural beneficence, David Barash claimed, 'Evolutionary biology is quite clear that "what's in it for me?" is an ancient refrain for all life' (Barash, 1977, p. 167). And the prominent biologist George Williams provocatively viewed 'natural selection as a process for maximizing selfishness' (Williams, 1988).

Such claims were legion during a prior era of evolutionary popularization. In response, metaphorically ascribing 'selfishness' to behaviour that merely introduces genes into the next generation – often through investment in others at cost of pain and life of the actor – has been appropriately criticized as illegitimate. But aside from this interpretative critique there remains the empirical question of whether there is *meaningful directionality* in evolution and, if so, whether it is hospitable to cooperation or dominated by conflict.

Views about progress in evolutionary history have waxed and waned since Darwin. There is evidence that in many respects evolution does exhibit trends that might be described as progressive. Features like taxonomic diversity, body size, homeostatic precision, sensory acuity and parental care increase on average across time. But, in addition, we now recognize a series of what are called *major evolutionary transitions* involving a trajectory that is in tension with the 'red in tooth and claw' view of nature. As life has moved from prokaryotic (very simple) cells to complex eukaryotic cells, from single-celled to multicellular organisms, from asexual to sexual reproducers, from solitary to (eu)social organisms, we see an ongoing theme. Each of these transitions reformulates, in a sense, a previously autonomous individual into a new, cooperatively interdependent unit with emergent levels of enhanced functionality. Studies in major transitions seem to validate a proposal made by Michael Polanyi a generation ago: 'the evolutionary sequence gains a new and deeper significance. We can recognize then a strictly defined progression rising from the inanimate level to ever higher additional principles of life' (Polanyi, 1968). Reflecting on this 'progressive' series of evolutionary transitions, Michod concludes, 'Cooperation is now seen as a primary creative force behind greater levels of complexity and organization in biology' (Michod, 2000, p. xi).

What these biological data show is surely not that the world or evolutionary processes contain no natural evil, but that evolutionary theory does not radically recast the ancient issue. The extent, role and dominance of what we view as natural evil have been over-simplified if not overtly mischaracterized by previous, nihilistic exegetes of evolution. However, this still leaves the question of why God's mode of creating allows for evolutionary evil at all.

Philosophical considerations: why does God allow evil?

But let's now turn to the philosophical challenges that are raised by the evil that seems to attend the evolutionary process. The phrase 'the problem of evil' is ordinarily taken to refer to one or another argument that reasons from the existence of evil to a conclusion that there is no God. The most widely discussed and defended version of that argument in contemporary philosophy can be stated as follows:

1 There exist instances of intense suffering which an omnipotent, omniscient being could have prevented without thereby losing some greater good or permitting some equally bad or worse evil.
2 An omniscient, wholly good being would prevent the occurrence of any intense suffering it could, unless it could not do so without

thereby losing some greater good or permitting some equally bad or worse evil.

3 There does not exist an omnipotent, omniscient, wholly good being.

The key premise in this argument is (1), since that is the premise that claims there are evils that exist for no good reason. It is the existence of evils of this sort, commonly called *gratuitous evils*, that raise the problem. So, if one wants to resist this argument, what tack should be taken? What is needed in order to undermine premise (1)?

The short answer is: the theist needs good reason to reject the truth of (1). But what exactly does that mean? One might think it means that we need to have some story about why God permits evils to occur, and which thereby allow for greater goods to emerge. But in fact rejecting (1) does not require that we meet such a high standard. All we need to be able to produce to reasonably reject (1) is an explanation of evil that is true *for all we know*.

Here is an analogy: if someone tells a man that his wife was seen holding hands with a handsome man in a romantic restaurant, should he conclude that his wife was being unfaithful? Not necessarily. If the man believes his wife is faithful, he could quite reasonably think that, perhaps, his wife was with her brother or cousin, comforting him after learning some bad news. He can reject the view that his wife is unfaithful based on this evidence since, *for all he knows*, there are all sorts of ways to explain this evidence.

When it comes to the problem of evil, the situation is analogous. If someone presents me with examples of evil which, as far as he or she can tell, occur for no good reason, from which the person concludes that there is no God, I can rationally resist that inference if I have an alternative explanation of that evil which is good, and is true for all I know.

We still need to say something about what a 'good explanation' of evil would look like. But here there is not much disagreement among experts. An explanation is good if it makes it clear that the evil that is permitted is a necessary condition for the occurrence of an outweighing good.

Poor explanations for the existence of evil: the fall, freedom and evolution

There are a number of explanations that have been offered for the pain, suffering, extinction, death, predation, and so on that seem instrumental to the evolutionary process. Some of these explanations, many of which enjoy widespread support, do not satisfy even the minimal demands as set out above. Before we look at some potentially successful responses to the problem, it will be useful and instructive to look at some unsuccessful responses. This will not only help to make clear what a successful explanation does look like, but will also help prevent us from latching on to explanations that are simply not up to the task.

The first explanation is the one that has probably enjoyed the greatest popularity in the history of Christian theology, namely, the explanation rooted in the fall of Adam and Eve. On that explanation, the moral wrongdoing, especially of Adam, unleashed cataclysmic consequences, not only for Adam and, more generally, the relationship between God and humanity, but also for the physical cosmos. Christians often point to a passage in the biblical text such as Isaiah 24.4–6 or Romans 8.19–22 as evidence for this view.

There are, however, two serious problems with the fall as an explanation of evolutionary evil. The first is that the vast majority of the evil in evolutionary history pre-dates the fall and thus seemingly cannot be explained by it. However, even if that issue could be overcome, the fall explanation faces a different worry not often addressed by its defenders. If the fall were to have these catastrophic consequences for the physical cosmos, this could only occur because God initially set up the creation so that moral evils of this sort would have such natural consequences. But that, all on its own, seems to be a substantial flaw in the design of the cosmos. Why, we might fairly wonder, would God constitute the physical cosmos in such a way that innocent non-human animals could become the innocent and unwitting victims of the wrongdoing of Adam and Eve? This seems to imply that the creation displayed a kind of *fragility* which made it liable to fracture and decay on a hair trigger.

There are variants of the fall explanations that would not be susceptible to this sort of worry. Gregory Boyd, for example, has argued that it is not the fall of Adam and Eve that wreaks havoc on the creation, but the prehistoric fall of Satan. This view is not susceptible to the worry about evolutionary evil pre-dating the cause of that evil. It might also escape the fragility objection because the natural evil that results from the satanic fall might be due, not to a built-in fragility which fractures the integrity of nature when Satan falls, but rather to specific acts of moral wrongdoing by Satan (e.g. tinkering with the genome in ways that promote suboptimal design).

A second, increasingly popular, explanation is that natural evil is not caused by moral evil, but can nonetheless be explained in a somewhat analogous way. On this view, creation itself, like the human beings it contains, has a 'freedom to wander' which necessarily leaves open the possibility that it will wander in directions that involve natural evil. John Polkinghorne defends this view. Polkinghorne says,

> It is the nature of cells that they will mutate, sometimes producing new forms of life, sometimes grievous disabilities, sometimes cancers . . . That these things are so is not gratuitous or due to divine oversight or indifference. They are the necessary cost of a creation given by its Creator the freedom to be itself. Not all that happens is in accordance with God's will because God has stood back, making metaphysical room for creaturely action. (Polkinghorne, 2003, p. 13)

However, this explanation falls short for at least three reasons. First, despite the label, non-conscious creation is not free at all, but rather, at best, contingent or unpredictable in its behaviour. We might be tempted to think of the contingency or unpredictability as a sign that nature as a whole is also free, but this would be to fall prey to an objectionable (even if common) form of anthropomorphism.

Second, even if the cosmos were free to wander, why did God allow it to contain as much evil as it does, persisting for so long? A universe created fully formed, in just the way hypothesized by so-called 'young-Earth creationists', could be equally 'free to wander'. But it would not include the massive suffering that is in fact part of our current evolutionary heritage. This gives us reason to doubt that this explanation satisfies the demand that the permitted evil be a 'necessary condition' for securing the outweighing good.

Finally, even if we can resolve these two problems, we are still left with the problem that the good of a cosmos that is 'free to wander' does not outweigh the resulting evil: all of the billions of organisms who suffer and die horribly. Is the good of a creation 'free to wander' worth the price? It seems not. And if not, this explanation fails to satisfy the other demand we require in a good explanation, namely, that the good that results from permitting the evil is 'an outweighing good'.

The third and final explanation that we shall consider here has been developed most vigorously by the biologist Francisco Ayala. Ayala claims that, rather than exacerbating the problem of evil, evolution solves the problem of evil, since the burden of blame for that evil falls not on God, but rather on the impersonal process of evolution itself. Ayala (2007, p. 159) explains the view as follows:

> A major burden was removed from the shoulders of believers when convincing evidence was advanced that the design of organisms need not be attributed to the immediate agency of the Creator, but rather is an outcome of natural processes.

On this view, God is relieved of responsibility, allowing theists to 'acknowledge Darwin's revolution and accept natural selection as the process that accounts for the design of organisms, as well as for the dysfunctions, oddities, cruelties, and sadism that pervade the world of life' (Ayala, 2007, p. 160).

This explanation is unsatisfying for numerous reasons. First, it does not meet the condition set out earlier for a good explanation of evil, namely, it does not show the permission of evil to be necessary for securing a greater good. What is the greater good? And why would permitting evolutionary evil be necessary to secure it? We are not told.

Second, this explanation merely transfers the problem up one level (or perhaps better: down one level). For, while God is not the direct cause of evil, God is still the remote cause, and thus no less responsible. We can see the problem clearly by considering an analogy. Wanting not to be responsible for driving when drunk and getting in an accident where passengers are drunk, one might choose not to drive when drunk. But it is no better to put one's passengers in a car with a known inebriated chauffeur! On Ayala's view, God has done just that with evolution.

Good explanations for the existence of evil: neo-Cartesianism, embodied intentionality and chaos-to-order

So what would a good explanation of evil look like? Below we will look very quickly at three possibilities.

When considering the problem of animal pain, suffering and death, one question that arises but is not often addressed is this: what is animal consciousness like, and how would we know? This is obviously an important question, since what most fundamentally makes animal pain and suffering a bad thing is the conscious character of the mental states of pain and suffering. If animal consciousness has an entirely different qualitative character, the extent to which it is different might change the extent to which those states count as evil.

Neo-Cartesianism

Scholars who investigate consciousness have described a number of theories that, if true, would have important consequences for how to think about evolutionary evil. The first is a view we might call *neo-Cartesianism*. On this view animals have consciousness of their surroundings – a consciousness that allows them to successfully navigate, catch prey, avoid predators, and so on. But those mental states are intrinsically very different from ours because having that sort of consciousness does not 'feel like' anything to them (this condition of having mental states 'feel like' something is usually called *phenomenal consciousness*). It is possible that animal mental states are like this for all we know. And if they are indeed like that, then, while animals may display behaviour that is similar to our behaviour when we are in pain, their mental states don't 'feel badly' to them. Not surprisingly, the neo-Cartesian view is not, to put it mildly, very popular. Indeed, most people think that the view *can* and *must* be rejected in the light of other things they know, though others disagree.

Another view of consciousness is based on a theory according to which mental states have a phenomenal character when those mental states are the object of another mental state called a *higher-order thought* (or HOT). So, on this view, when I step on a drawing pin, this induces a mental state in me; and when I direct a higher-order thought at that mental state, then, and only then, does the mental state 'feel' like something. Advocates of this view think that this sort of view can explain certain strange phenomena like *blindsight*. In blindsight, patients claim that they cannot see anything at all, but they can also perform behaviours that require sight (like navigating through a hallway filled with objects, without any assistance). HOT theorists argue that blindsight patients have visual mental states, but those states don't 'feel like' anything to them, and so they do not have any 'awareness' that they have them.

If animals lack higher-order thoughts, and this theory of consciousness is right, then animals have no conscious awareness of their pains, even though they have pain states. They have, in the case of pain, something analogous to blindsight.

What if we reject these views of consciousness and admit that animals have painful mental states, and have a phenomenal conscious awareness of them? In that case, animal mental states are very much like ours. If that were the case, would that entail that animal pain and suffering is morally bad? The answer is: not yet. The reason is that we know that even human beings with phenomenally conscious pain states sometimes regard those states as not being unpleasant. In the 1930s and 1940s Dr Walter Freeman developed a procedure that came to be known as *lobotomy*, in which he destroyed portions of patients' prefrontal cortex. The procedure was sometimes used with patients with chronic pain. Afterwards, these patients would sometimes express relief, but not because the pain was gone. Rather, they would report that the pain was still there, and just as intense, but that they did not care about it any more. More recent work in neuroscience shows that pain is mediated by two pathways: one that detects the injurious stimuli, and one that moderates the level at which we care about it.

Since almost all other animals lack a part of the brain that is associated with 'caring about' our mental states, one possibility that presents itself is this: animals have conscious pain states, but having them does not bother them. Of course, animals would still display all the behaviours that *we* associate with unpleasant pain. Presumably those behaviours evolved because they are adaptive behaviours to display when we are injured, not because they necessarily signal discomfort. So we are left to wonder: is that what animal mental states are like when it comes to pain? And the answer is: we do not know. But, if any of these proposals are correct, it would seem that animal pain and suffering is not a problem to be solved in the first place. And the problem of animal pain and suffering thus evaporates.

Embodied intentionality

But let's suppose all of these theories are wrong. What alternatives would there be for explaining animal pain and suffering in the context of the evolutionary problem of evil? Some philosophers have argued that animal pain and suffering is justified because it is a necessary condition for a greater good that accrues to the animals themselves. On one proposal, animal pain and suffering permits animals to engage in morally significant or even heroic acts by putting themselves at risk. The mother who plays the decoy to protect her offspring from a predator does something of moral significance by putting herself at risk of suffering or death. Another proposal holds that the possibility of pain and suffering gives animals the opportunity to develop something like a moral character that enhances their flourishing either in this life or in heaven in the presence of God. However, these proposals seem, like Polkinghorne's above, objectionably anthropomorphic. And there are additional problems. For example, mothers can play decoy and incur a risk of loss even without pain and suffering, since they can be at risk of losing their very lives.

There is, however, another way of developing theories along these lines. If animals act on intentions – if they *intend* to do things when they act – there must be some mechanism that allows them to balance their desires to achieve the intended end, and the risks that the intended acts pose to their bodily integrity. There is good reason to think that the only sort of tool that can provide the proper counterbalance is pain. We should be clear that we are not saying that in order for animals to avoid injury it is necessary that they be 'wired up' with a system that alarms them to injury. There are all sorts of ways that animals could be wired to cause them to recoil from environmental dangers. What we are saying is that, when (if ever) animals *act on intentions*, they need a countervailing source of motivation that can deter them from pursuing those intentions, when those intentions also involve a risk to their bodily integrity.

Of course one might still wonder why that countervailing motivation has to be *pain*. And, in fact, we have some answers to this question. Earlier in the twentieth century the physician Dr Paul Brand worked with patients suffering from Hanson's disease, a disease that causes loss of pain in one's limbs and digits. He tried a variety of ways to replace the lost sensations that indicate that we have bodies that are incurring physical harm. He and his patients tried a variety of substitute mechanisms, but only pain would work – and that had to be pain that could not be overridden. If Brand is right, pain is not just *one way* to keep intentional agents from injuring themselves; it is either the only way or the most effective way.

Chaos-to-order via law-like regularity

There is, however, a third account of why God permits evolutionary evil in the form of animal pain and suffering, which might be developed as follows. One common explanation for evil invoked by theists holds that natural evil occurs as an inevitable by-product of a physical universe governed by laws that are rarely overridden by miracles. A universe like that is bound to include instances where the laws of nature 'collide' as it were, and generate events that we call natural evil. We might call this the *law-like regularity* explanation for natural evil.

What goods might law-like regularity be necessary to secure? There are many possibilities; here is just one. In order for anyone to be able to 'intend to perform an action', it must be the case that the world is law-like regular. For me to intend to throw a cricket ball there need to be regular laws of motion that allow my moving hand and arm to propel the ball through the air. Without that regularity, throwing would not happen, I would not learn that it does happen, and I could not intend to make it happen. So, in short, law-like regularity is a necessary condition for the good of intentional (and thus free) action.

Some Christians have also argued that law-like regularity is a *good* feature of nature. Augustine, Leibniz, the French Catholic philosopher Malebranche and many others famously argued that a regular 'clock-work' universe is an intrinsically good type of creation.

But what does law-like regularity have to do with animal pain and suffering in the context of evolution? It might require lots of 'collisions' and thus lots of natural evil, but it is not clear why it requires evolutionary evil. After all, the world can be perfectly regular and still be the product of a six-day creation a mere 10,000 years ago. No evolutionary process is required or entailed by law-like regularity.

However, we can supplement law-like regularity with another feature which *would* necessitate animal suffering. What we need to close the loop here is an explanation of why the good (or goods) at which law-like regularity aims is better realized in a universe that *proceeds from chaos to order by law-like means*. The notion that a universe that brings about order over time is a good thing has been defended by a number of Christians throughout the ages.

Still, even if this 'chaos-to-order-via-law-like-means' is a good, why must the creatures that preceded human beings experience pain and suffering? Perhaps the answer is that those precursors were intentional agents who needed it for protection, as we saw earlier. Or perhaps it is because evolution required conscious precursors to human beings as a natural rung in the evolutionary ladder that leads to human beings.

Of course, even if we agree that chaos-to-order-via-law-like-means is a good, and that it necessarily leads to certain sorts of evil, there is still the question we faced before: is it an outweighing good – a good that justifies permitting so much evil? We expect many will say 'no'. But the verdict here is not transparently clear.

Note that in all of this we are not proposing a *solution* to the problem of evolutionary evil. Rather, we are pointing to some possible solutions that, on the surface, seem to satisfy the criteria we set out at the start. Are all of these solutions true? The answer is: 'no'. And they cannot be since they are, in fact, inconsistent with one another. One postulates that animals do not feel pain, while another posits that they must! The key here is that the explanations we have offered are all serious candidates in that they explain why the evils in question are necessary for achieving some outweighing good, and such that we are not entitled to rule them out entirely, given the things that we know. That is a low standard but, as we have seen, a nonetheless hard task.

8

Is there more to life than genes?

PAULINE RUDD

The body

Certainly the body isn't one part but many. If the foot says, 'I'm not part of the body because I'm not a hand', does that mean it's not part of the body? If the ear says, 'I'm not part of the body because I'm not an eye', does that mean it's not part of the body? If the whole body were an eye, what would happen to the hearing? And if the whole body were an ear, what would happen to the sense of smell? But as it is, God has placed each one of the parts in the body just like he wanted. If all were one and the same body part, what would happen to the body? But as it is, there are many parts but one body. So the eye can't say to the hand, 'I don't need you', or in turn, the head can't say to the feet, 'I don't need you'. Instead, the parts of the body that people think are the weakest are the most necessary.

(1 Corinthians 12.14–22, CEB)

In this passage, St Paul is referring to parts of the body that we can see, but equally important are the millions of molecular machines and processes that we cannot see but nevertheless sustain our every waking moment. These are no less a part of the body, and it makes no sense for the neurons to say to the heart, 'I don't need you'; neither can proteins say to the sugars or lipids, 'I don't need you', nor can the genes say to the proteins, 'I don't need you.'

And so, continues St Paul (1 Corinthians 14.24–26, CEB):

God has put the body together . . . so that there won't be division in the body and so the parts might have mutual concern for each other. If one part suffers, all the parts suffer with it; if one part gets the glory, all the parts celebrate with it.

The body is a highly complex entity which is constructed in such a way that it can interact effectively with the environment on which our survival depends. We are part of a holistic system – the body cannot say to the atmosphere, 'I don't need you', and live; we cannot say to other people and other species, 'I don't need you', and continue to live.

If we wish to describe a human being or any other organism, it is hard to know where to begin, because there is no single beginning. We might try to begin chronologically, but everything, including the fertilization of the egg by a sperm, takes place in an environment that was prepared before the event. We might try to begin with an atom or the smallest component of a cell, but very quickly there would be so many options that a linear story would soon become impossible. The organization of a human being, or any other creature, does not have a single, simple hierarchy. There is no means of describing ourselves in a linear fashion, for there are countless starting points and hundreds of feedback loops that pass information around the body. This allows us to respond to our constantly changing environment and to engage in the cooperative behaviour that enables us to survive in a challenging world.

Our genes

In our search to discover ourselves, our *genes* are a good place to start, for they make each of us unique. They do not act unilaterally, and they do not constitute a blueprint for our bodies; however, they do contain information that represents potential and imposes constraints.

We each have our own particular version of the human genome inside our cells. From our 30,000 or so genes, the systems they operate and the environment that surrounds us within and without, we derive purpose, assume agency, transcend the limitations of our environment and accept constraints that we cannot change. To this extent our genes are indeed us; these we can claim are peculiarly our own.

So what are genes? The way in which geneticists describe genes is changing. They were once defined as units of inheritance. A more recent definition is that *genes are DNA-based units that can exert their effects on the organism in which they are located through RNA or protein products.* Genes are one of the three biological alphabets that provide the building blocks of the natural world. They are composed of combinations of four bases (adenine, thymine, guanine and cytosine) and are assembled on very long linear molecules of deoxy-ribonucleic acid. One strand of DNA with the four bases attached and its complementary partner combine to form the famous double helical structure. Genes are arranged on 23 pairs of chromosomes in humans (a chromosome pair is composed of two molecules of DNA). Some are bigger than others and they are numbered roughly in order of size. There is enough DNA in the human body to stretch to the Sun and back 600 times!

Genes and proteins

Genes can replicate themselves. Also, the DNA can be specifically unravelled to expose particular genes when they are needed to make *proteins*. The genes are copied to make messenger RNA (mRNA) that finds its way to the ribosomes located in a part of the cell called the endoplasmic reticulum. In the ribosome mRNA is edited and translated into proteins. These are linear molecules composed of *amino acids* linked together. Three bases code for one amino acid. Human beings make 21 different amino acids. The gene is 'read' from one end to the other and amino acids are linked together in the order specified by the base triplets to form the protein.

Genes can be altered, that is, mutated, when the order of the bases is changed. Mutations alter protein sequence and structure that may or may not alter the function of the protein, and might lead to new opportunities or sometimes to disease.

Many genes code for proteins that have multiple functions; all cells contain all of them, but genes are only expressed in specific temporal and spatial locations. They are carefully controlled switches that allow the growth, repair and differentiation of the body. Very few genes give rise to a single consequence or even act alone. For example, 100 genes are related to height, and of course nutrition also plays a role; determining height is a very complex process and it is no wonder that we cannot add one cubit to our stature by worrying about it!

Genes respond to signals from their environment telling them that the protein they code for is needed. They are devices for extracting information from the environment. To a large extent genes have the potential to determine the organization of their own body within an awe-inspiring complexity. Just as music has an existence of its own but requires an instrument to be realized, genes need a body in which to be expressed. However, a person may have the genetic background to be a great violinist, but if they are never given an instrument or do not practise, their gift will never be developed. Some things are less about genes and more about opportunity.

Genes have interchangeable parts that code for important functions. During evolution, when a protein had a useful function, the key parts of it were retained; that is why we have large numbers of genes in common with fruit flies, the roundish flat worm and the stickleback. For example, epidermal growth factor, which has a crucial role in cell growth and proliferation, is present in hundreds of species from lampreys to chimpanzees. The gene may mutate a little and the protein sequence will change, but if that change is so great that it means an essential protein cannot function adequately, then the mutation will be lethal and will not persist in the species.

Genetic mutation and diversity

Genes can be altered (mutated). This happens, for example, when one base changes, perhaps because of miscopying. This in turn changes the sequence of amino acids and thus alters protein structure. This may do nothing, but if it is a sensitive part of the protein it can have a dramatic effect. It might give the protein an alternative useful function, but more often it builds an inactive protein that can cause serious disease – as in the family of disorders known as 'congenital disorders of glycosylation'. In one case, the mutation in a single critical gene, Mgat 2, which makes an enzyme involved in sugar processing, leads to a heavily compromised individual who has a seriously impaired immune system.

Another genetic variation, this time of the HLA-B gene, HLA-B53, which some people carry, makes a particular immune molecule that protects them against severe malaria. So the sequences of our personal genomes can give us protection against some pathogens but be risk factors for others.

Genetic diversity not only makes each of us unique; it also helps to ensure the survival of our species. Darwin's finches on the Galapagos Islands have been studied for many years. There are several subspecies with different kinds of beaks that are specialized for different types of food. Some birds do better in dry seasons, some in the wet. Whatever the conditions, one or other subspecies will be adapted to survive.

Our individual genetic differences have enabled us to withstand plagues, and also to specialize and divide our tasks within a community, liberating time for creativity and imagination – we could never have progressed if we could not share skills. Economies arose because of exchange of objects and agreements to specialize. Our culture did not shape us as a species; rather, it developed because of our collective abilities and because of our development of language and technologies and our expanding brains.

Our brains

Our brains are the most complicated organ we have, and we are only just beginning to understand its genetics, its proteins and sugars, and how it works. We do know that many of our activities are not normally exercised under conscious control. The brain is smart – there are about 40 firings per second at the conscious level compared with 40 million at the subconscious level. This allows the brain to give maximum attention to things that need conscious, intelligent, executive decisions.

The brain is plastic and continues to develop and adapt from three weeks after conception until the end of our lives. It responds to experience, forming new pathways; for example, the prefrontal cortex is not fully developed until about the age of 25.

We have outlined a path that takes us from genes to the most complex organ on the planet, the human brain. We have seen that the cell can take messages from its environment, and transcribe genes to proteins to produce fully functional molecules. At each step we can see that small components, such as bases, give rise to entities that are more complex and diverse (the genes and proteins), and that these complex and diverse entities can themselves be viewed as simple and unified when we analyse what is happening at the next level of complexity (the cells, tissues and organs, and eventually the whole person). But the big question is: who or what is ultimately in control?

Life: going beyond genetic determinism – we are more than our genes

One thing we have learned in recent decades is that life is infinitely more intricate than we imagined. New technologies have enabled us to open up the field of biology to explore the chemistry that underpins it. We can now appreciate that we are made up of thousands of dynamic systems, many completely outside our conscious control. It seems that there is nothing ultimately in control, no single thing directing and micromanaging us. Each small part faithfully carries out its role, unaware of the place it has in the big picture until finally we reach the level of consciousness. Even then it seems that this is a cooperative, integrated process with many inputs from inside and outside our bodies. So what initiates action?

It used to be axiomatic that *reductionism* was the scientific method that would allow us to describe and control everything, and indeed it is still an essential part of scientific investigation. But it is not enough; in modern biology we have reached the limits of our ability to rebuild the whole picture from analysing its parts. As we begin to appreciate the level of detail that underlies the simplest of operations in our bodies, we need all the power of modern bioinformatics to work out how to assemble large amounts of non-linear information.

Trying to deal with complexity and emerging properties is a major challenge that defies a simplistic view of the world. Many things do not fall neatly into boxes and we are currently trying to understand the *tipping points* that lead to committed action such as the differentiation of a stem cell. We need to come to terms with the reality that everything is dynamic; many entities have several options when it comes to activity and we need to visualize thousands of interactive pathways, preferably simultaneously.

At the biochemical level, we are certainly more than our genes. Genetic determinism is not a general feature of our individual genomes, for the external and internal environments are the backdrop that triggers gene expression. Our inner worlds matter too, for our brains have given us the possibility of imagining what we cannot see and have not experienced. It is remarkable that our material brains can engage with abstract ideas that come from other people and with emotions that are non-material, and then take action. As many sages and ordinary mortals have testified, it is also entirely possible for us to experience insights that did not arise from rational conscious deduction and, at the time, were beyond description.

The challenge now is to discover how we can relate all the detailed information we have about our physical bodies to the deepest yearnings and insights of the human spirit. Science is an incredibly effective way of understanding the world, but alone it is not enough. Having an intellectual understanding of causes is not the same thing as dealing with the consequences.

Science may provide a rationale for the basis of our ethics, our relationships with others, and our deepest experiences of love, joy, beauty and hope. But understanding how we work is not the same as understanding how we should live. Science, like nature, is ethically neutral; we have to think out the rest for ourselves.

To be more specific, if we reject predestination (which would be imposed by absolute genetic determinism) and if we no longer subscribe to the concept of a god who controls our lives through divine intervention, then we ourselves become responsible for determining the purpose of our lives, and that means taking into account the consequences of our personal genomes, those of others and our environment. Even if it did turn out that we are genetically predetermined or controlled like puppets by some god out there, it would not really help, because few of us really believe that we have no personal freedom or responsibility. Our perception of ourselves as having some level of self-determination and agency might just be an illusion, I suppose, but the alternative, that we have neither of these, would seem to fly in the face of common sense.

In any case, this is in fact how we live, for regardless of our certain knowledge that our lives are finite, that the Earth will eventually be subsumed and in the end all that we know will no longer exist, we continue to imbue our existence with meaning. This demanding approach to life, which in the long term appears to be rather irrational, seems to need addressing. To approach this question, I believe we need more than a mechanistic view of our world, indispensable though that is for many purposes. If we are to make the complex choices that confront us with sound judgement, then we need to complement the magnificence of our modern scientific insights with experience gained from other windows on the world. These include art, music, literature, and the great philosophies and religions of the world.

Combining imagination and rationality:
our experience of God

The scientific enterprise itself is enabled when imagination and insight are combined with rational thinking. Bringing ideas from our unconscious 'knowing' to the conscious mind where they can be evaluated and codified is a practice that every scientist, artist, mathematician and philosopher will recognize. Scientists use model systems and equations to express ideas that begin in the imagination. Religious thinkers use poetry, ritual, symbol and myth to express theirs. Scientists test their models in experiments. Religious leaders and philosophers develop internally consistent doctrines and dogmas. Those who follow a spiritual path test their inner experience in their own lives.

For millennia people have projected their profound internal awareness on to higher beings of all kinds in attempts to link the sacred with the secular. Many reasons have been offered by anthropologists for this practice. However, God in essence is unknowable – if we claim to know God, then 'it is not God that we know', said St Augustine. Any description will be limited, for even Moses could not gaze upon the face of God.

Perhaps what we are doing is describing a profound experience, projecting on to it attributes that are the best that we know, such as justice, mercy, steadfast love, truth, compassion, charity, harmony, forgiveness and redemption, eternal life and even ultimate reality. This experience is not a reality in the scientific material sense that it can be measured, but it is powerful. It can profoundly affect human behaviour. It is hard to define precisely, because by engaging with it we personalize it. We can add our own insights to the whole and draw on the insights of others in our personal quests for meaning, purpose and courage.

Our experience of God is a deep one. When we experience something we choose to name God, we may not be able to prove that we are communicating with any external reality, yet we can say that we understand that the universe beyond us is not empty of consciousness – deep communicates with deep. Or as Newman wrote, '*Cor ad cor loquitur*', which may be translated as 'Heart speaks unto heart' (Newman, 2014).

Many people pray unceasingly that the great heart of our own heart will still hold our vision whatever assails us. But note now how Newman's language and mine are becoming less scientific and less precise than the earlier part of this chapter. This is important, because inspirational experience is particular to the individual. Furthermore, there needs to be space and flexibility in language, so we can relate to insights from our own tradition and from others.

How can these insights inform and enrich our lives? The immense creative burst that is the universe that has given birth to us is displayed in endless forms which inspire and delight us. We can only marvel at the vast power of matter and energy that combined to form everything from the stellar displays in the night sky to the unseen molecular machines that provide our cells with energy. Our universe is the manifestation of this creative force that in most religions provides a foundation for the concept of God. Humans are also filled with creative energy and we bring this to birth in any number of forms, including art, music, drama, science and literature. Innate biological systems control the mechanisms that sustain life; our cultural inheritance helps us to make the most of our environment; yet a truly original creative discovery by an individual is novel – it grows and blossoms but is not preprogrammed or learned. Once we become aware of new insights, our impulse is to create something that will allow us to articulate our inner experience, so that we can integrate it into our own lives and share it with others. This is not an easy process, and we are continually challenged to bring our informed and intuitive inner knowing to a reality that we can integrate into other forms of art or language, to use as a platform of consolidated ideas from which to step further into the unknown.

There is no single entity in control of us. Not our genes, not our brains, not God. Every cell in our body has a completed copy of our genome and is sensitive to its own internal and external environment. There is no top-down command system. Cells work in partnership with other cells around them. As we mature, we recognize that we can be in a co-creative partnership with God. Attempts at total control, exercised either by ourselves or by God, may satisfy our desire for security, but they don't really fit with our observance and experience of nature, or indeed with experience of ourselves at the level of intentional action. It appears, instead, that there is openness, a range of possibilities within nature. If either we envisage God as someone who designed everything once and for all at the beginning of creation, or we envisage ourselves as the ultimate controllers of the natural world through science, then we need to take care that we do not preclude the possibility of creation itself being creative as it responds in harmony with changes to itself. Perhaps our understanding of God's involvement with us is such that together we can attain real novelty, contingency and opportunity that preserve the integrity of life in the process.

Challenges of integrated biology: sea changes in our worldview

Certainty and uncertainty

We can analyse the building blocks of nature with a greater or lesser degree of certainty. Interestingly, ever since Descartes tried to distinguish between science and magic by setting out to believe only what he himself could prove for certain, philosophers of science have pointed out that he was really addressing the wrong question. Science does not deal with certainty but with knowledge (knowing things) and probability. Indeed, many facts that we know for certain, such as that the world is older than five minutes, are rather uninteresting and banal.

In everyday life, as well as in science, we rarely prove anything to be true beyond doubt, but we garner enough supportive evidence to risk making a decision. If we are to make progress or avoid disaster, we need to be able to assess the available evidence and then make a judgement in time to affect a situation. We have not waited to ban smoking until we have incontrovertible evidence that it kills us, and hopefully we will not wait until global warming overtakes us before we act. And if you think about fractals and the question of absolute measurement for a moment, it's clear that we can never know exactly the length of the coastline of the UK. If we are to build a house we have to deal every day with pragmatic approximations. Like being a parent, rather than being perfect (whatever that means), we need to be good enough to achieve our goals.

In almost every field, from astronomy to meteorology, modern science has come to the realization that the natural world works by making new beginnings that arise from uncertainty. We have moved away from the idea of a deterministic universe where everything is as predictable as balls moving on a snooker table to a world of dynamic flexibility and ever-increasing possibilities. The need to address the big question 'What is it all for?', both individually and collectively, has never been greater.

Such a dynamic, less predictable universe in which we now live is much harder to describe, and it is clear that in many fields we are only just beginning to develop the technical and bioinformatics tools that will enable us to deal with complexity, emergent properties and big data sets. We need to do this so that we can in a meaningful way not only measure, but represent and describe, the dynamic complex processes that are going on simultaneously in an organism.

Big data

One of the latest buzzwords that is often misused is *big data*. This concept is not so much about acquiring large amounts of data as about collecting information-rich data that will change the way we think about the world and our place in it.

The Victorian worldview was transformed when ordinary people looked at the skies with telescopes. Compared with what we know today, the amount of data they collected was not huge. However, it inspired not only astronomers, but artists like van Gogh who painted his famous night sky with huge stars hovering over the landscape, musicians like Holst who captured the majesty of the planets with *The Planets Suite*, and H. G. Wells and C. S. Lewis who wrote science fiction novels about life on other worlds. The realization that the heavens are far bigger and that the Earth is far more insignificant than they ever imagined was deeply unsettling. The imaginations of the artists were vital in enabling people to accommodate this new view of the world and our place within it.

Tipping points

Our modern insights into the dynamic complexity of living organisms are in many ways as unsettling as the transformation of the Victorian world by the telescope and microscope. To achieve a deeper under-standing of our analytical data and its impact on biology we need to know about the biochemistry of organisms, and to do this we need to link different facets of the biological alphabets together. There is a long way to go yet. We are at the foothills, for at the moment we are mostly still looking at associations and correlations, which do not equal causation. However, it is clear that neither the gene nor the primary amino acid sequence is sufficient to define the structure and function of a protein. Epigenetic modifications, the transcriptome, metabolome and lipidome, the glycome and the local environment are just a few aspects that come between the gene and the expression of the functional protein it codes for.

All of these different systems are a means of generating subsets of a protein that can diversify its structure, function or location, thus multiplying the potential roles of a single gene product. Moreover, this type of diversification can be attained without recourse to genetic modification of the gene for the target protein, so it provides rapid ways of responding to change.

If genomes are the beginning, then glycomes are surely the end, for the addition of sugars to proteins is one of the final steps in the making of a glycoprotein. In the long path from gene to glycoprotein any number of interactions may be suboptimal. In his book, Malcolm Gladwell defines a *tipping point* as 'the moment of critical mass, the threshold, the boiling point' (Gladwell, 2002, p. 12).

A big question for understanding disease is, 'How many and which suboptimal processes are needed to give rise to a disease?' Are there some steps that can never be compensated for which make disease inevitable? Are there others that can be tolerated because the cells can find alternative pathways? How suboptimal does a process have to be before it becomes lethal? What are the critical features of genes, proteins and sugars that will allow us to probe this question?

Modern technologies and detailed data analysis

It is clear that modern technologies are enabling us to perform very detailed analyses of all kinds of molecule. However, as we develop better and faster instruments, the amount of data becomes over-whelming and we are in danger of diluting the signal with noise as techniques become more and more sensitive. It is informative to listen to astronomers who have been creative in working out ways to store and organize huge amounts of data. Maybe, even though 85 per cent of the organic mass on the plant is carbohydrate we do not have as much difficulty as those who elect to map the heavens!

In 2001 the analysis of the human genome mark 1 was completed. However, it was not until 2006 that the field of genomics really took off, since it needed the development of high-throughput, chip-based, dense, genome-wide scans using hundreds of thousands of common single nucleotide polymorphism (SNP) markers. Other technologies are poised to advance our need to align glycomics with proteomics, lipidomics, genomics and epigenetics to bring analytics into a systems biology context.

The importance of detailed analysis in arriving at an integrated description of a big picture of the world is not new. Lucretius, in his poem *De Rerum Natura* (The Nature of Things) at the beginning of the first century BC, wrote:

> Moreover, it is vital in what order I array
> The different letters that make up my lines, in what position,
> Because the sky, the sea, the soil, the streams, the shining sun
> Are drawn from a single pool of letters, and one alphabet
> Spells barley, bushes, beasts, words not identical, and yet
> With certain letters shared in common, for what really matters,
> What makes a world of difference, is the arrangement of the letters.
> The same goes for the physical, for when you rearrange
> Atoms, their order, shapes and motions, then you also change
> What they compose.   (Lucretius, 2007)

Why are we doing all of this?

I was walking on the beach in Asilomar a while ago, thinking of the seemingly impossible amount of work I had committed myself to complete in a very short time and wondering why on earth I was attempting to do it. To avoid the question, I began to ponder the unimaginable amount of time that the sea has been pounding these ancient rocks and to think of the Ohlone or Costanoan people who have lived around Monterey for many centuries. It was then that I was reminded of the first poem I ever learned and which continues to inspire me as a positive answer to this eternal question. Hiawatha was a Mohawk chief in upstate New York who is credited with joining together five tribes to form the Iroquois confederacy. In a prelude to the legendary discovery of maize, the staple food of America, Longfellow writes:

> And, in accents like the sighing
> Of the South-Wind in the tree-tops,
> Said Mondamin, 'Hiawatha!
> All your prayers are heard in heaven,
> For you pray not like the others;
> Not for greater skill in hunting,
> Not for greater craft in fishing,
> Not for triumph in the battle,
> Nor renown among the warriors,
> But for profit of the people,

For advantage of the nations.
From the Master of Life descending,
I, the friend of man, Mondamin,
Come to warn you and instruct you,
How by struggle and by labour
You shall gain what you have prayed for.
Rise up from your bed of branches,
Rise, O youth, and wrestle with me!'   (Longfellow, 2010)

Communication

Cells form part of organisms and systems. They are present in the maize plant, the liver cells or the heart, each with a specific role. A cell is like a buzzing, overcrowded, well-organized city. It has compartments and means of transporting molecules from one to another. It has tracks and pathways to construct molecules, assemble molecular machines and deconstruct faulty products. It receives information from the extracellular environment and makes products in response to the signals. A cell *knows* what kind of cell it is, it produces its own housekeeping molecules, and it makes specific molecules for secretion or for displaying on the cell surface. It communicates within itself and with the environment and other cells beyond its surfaces.

Our view of the world is determined by our senses. We see the world through the lenses of our colour vision; we feel the size of things that are solid to our touch; we hear the sounds in our acoustic range; we smell and taste with receptors in our noses and mouths. The world as seen by a bee or a jellyfish would look very different. Reality is not all it seems.

We communicate with each other and the world around us through our senses and we construct a reality that enables us to navigate this physical world. There is a common basis to life, and so, depending on our level of empathy and imagination, we can in some measure communicate with animals and even plants and fish.

In the countryside we can be especially aware of how birds communicate with their own species and with others. We can see swallows and martins navigating the sky at speed but never colliding, or a magpie sitting on a branch dive-bombing and breaking up two fighting crows about to draw blood. The honey bird, the honey badger and human beings collaborate to raid the bees for the sweetness of their honey.

Words and actions can divide or unite us; they are culturally interpreted and we need to reach for our common humanity for unity. Common purpose unites us, and whatever the colour of our skin or the things we hold dear, in the beginning we all came out of Africa.

We are accountable for the way we choose to live, if not to our gods, then to our descendants.

For nature does not need us; we need nature. We do not have ultimate control, regardless of how well we can predict and understand its forces. Nature will always adapt. The question is, 'Will we?'

Conclusion

To summarize: human beings live in a glorious technicolour world in which we are required to deal decisively with the messiness of everyday life on a minute-by-minute basis, despite the complexity of the information that both drives and informs us. We continually integrate simple information into more complex systems. Our most basic decisions are guided by the possibilities and limitations imposed by not only our genes but many other molecules and the environment. The transcription factors that enable gene expression, the metabolic systems that supply our muscles with energy, the repair mechanisms that enable us to survive environmental insults, and our immune systems that combat disease operate largely without any conscious intervention by us, according to principles that we are increasingly understanding.

However, living a fulfilled human life is far more complicated than simply remaining alive. As individuals we need to become integrated into an ever more complex world so that we can find a niche where we can flourish physically and emotionally. Our cultural environment teaches us how to relate to others, and gives us the practical skills and learning that we need to deal with the complex world of work and family.

Yet most of us expect to achieve even more in the fleeting moment of consciousness that we are privileged to experience. Expressing our own particular creative ideas and responding to beauty, love, joy, sorrow, death and loss requires even more complex information and courage than our genes and culture alone can provide. This level of self-expression that is uniquely our own arises from a synthesis that takes place deep within the human psyche; it culminates in an awareness of an environment that we grasp first as tenuously as a dream, which, even as it slips through our fingers, we struggle to articulate so that it can be of lasting value to ourselves and to others.

So let us return to St Paul, in Philippians 4.8 (KJV):

> Finally, brethren, whatsoever things are true, whatsoever things are honest, whatsoever things are just, whatsoever things are pure, whatsoever things are lovely, whatsoever things of good report; if there be any virtue, and if there be any praise, think on these things.

Maybe in such an environment we will educate our genes.

Life can be lived with or without an in-depth knowledge of both science and religion, but how much richer it can be to explore the life-enhancing, complementary aspects of both.

9

Psychological science meets Christian faith

DAVID G. MYERS

I enthusiastically participate in (and report on) psychological science. And I am a person of active faith. So, I am sometimes asked, how do I reconcile these two worlds of thought within my one head?

My dance on the boundary of science and faith has entailed:

- explaining to people of faith the value of psychological science;
- comparing big ideas about human nature found in both psychological research and biblical and theological literatures;
- documenting some interesting links between faith and personal and social well-being;
- shining the light of science on some issues that people of faith debate.

The Christian mandate for doing science

As a person of faith I operate with two assumptions which, together, form the ground for doing free-spirited, truth-seeking science:

1 There is a God.
2 It's not us.

If, indeed, we humans have dignity but not deity – if we are finite, fallible creatures – then our surest conviction can be that some of our beliefs err. Thus, we had best hold our own untested beliefs tentatively and assess others' beliefs with open-minded scepticism. Moreover, when appropriate, we can use observation and experimentation to winnow truth from error.

Such faith-based humility and scepticism helped fuel the beginnings of modern science. This science-supportive attitude – which is espoused by my own 'Reformed and ever-reforming' Christian heritage – not only tolerates our participation in free-spirited scientific enquiry, but mandates it. The whole truth of God's creation cannot be discovered by introspectively searching our own finite minds.

So, we submit our tentative ideas to the test. If our ideas survive, so much the better for them. If they crash against a wall of evidence, so much the worse for them. So advised Moses (Deuteronomy 18.22, esv): 'When a prophet speaks in the name of the LORD, if the word does not come to pass or come true, that is a word that the LORD has not spoken.' Also, as Paul advised in his first letter to the Thessalonians (5.21, esv), 'Test everything; hold fast what is good.'

Such ever-reforming empiricism has many times changed my mind, leading me now to conclude that parenting practices have but modest effects on children's later personalities and intelligence; that crude-seeming electroconvulsive therapy can often relieve intractable depression; that the automatic unconscious mind dwarfs the conscious mind; that traumatic experiences rarely get repressed; and that sexual orientation is a natural, enduring disposition (not a moral choice).

Faith-supported scientific inquiry also has led me to *dis*believe certain spiritualist claims ranging from aura readings to out-of-body 'frequent flyer programmes'. If, for example, aura readers really can detect auras above a person's head, then they should be able to guess the person's location while seated behind a screen. If not (as seems the case), let's consider the claim discounted.

For Christians, the consistent failures to confirm such paranormal claims confirm the distinction between deity and humanity. We assume we are not little gods with powers of *omniscience* (reading minds, foretelling the future), *omnipresence* (travelling out of body) and *omnipotence* (levitating objects or eradicating tumours with our mental powers). As Isaiah 46.9 (niv) records, 'I am God, and there is none like me.'

So far, I have suggested that Christians in psychology feel called to explore God's human creation with a spirit of curiosity and humility. Believing, with John Calvin, that in everything we deal with God, we also feel called to worship God with our minds – through disciplined scientific inquiry – as we search God's world, seeking to discern its truths.

The psychology–religion interface: an overview

Beyond this, psychology and faith intersect in six additional ways, the latter three of which this chapter will illustrate.

1 *When teaching, writing, researching and practising psychology, we are mindful of our assumptions and values.* As psychology's Marxist, feminist and Christian critics have observed, the discipline is not value-neutral. When first drafting my psychology textbooks, I posted on my office door C. S. Lewis's reminder that 'We do not need more Christian books; we need more books by Christians about everything with Christian values built in.' When choosing to study and write about value-influenced topics such as evil, pride, prejudice, peacemaking, sexuality and altruism, we subtly and inevitably express our values.

2 *We apply psychological insights to the community of faith.* For those trained in seminary pastoral counselling programmes, this includes harnessing psychological therapies in caregiving. As a social psychologist, I have suggested how psychology's social influence principles might assist the creation of memorable, persuasive homilies and more effective evangelism.

3 *We study the psychology of religion.* Psychologists have studied various universal human phenomena, including sleep, sex, anger and hunger. Some 68 per cent of humans report that religion is 'important in their daily lives' (in a recent Gallup World Poll that my colleagues and I analysed – see below). So why not also put religious belief and behaviour under the psychological microscope?

4 *We compare psychological and religious understandings of human nature.* We ask how psychologists and theologians view the interplay between, for example, mind and body, belief and action, healthy self-esteem and toxic pride.

5 *We observe the apparent effects of religion.* Is religiosity associated with prejudice? Altruism? Human flourishing?

6 *We probe points of seeming tension between psychological science and faith.* What do experiments on illusory thinking and tests of intercessory prayer suggest about the efficacy and integrity of our prayers? What does research on sexual orientation and the human 'need to belong' imply for the church's stance on same-sex relationships and ordination?

Human nature in psychological and Christian perspectives

As Malcolm Jeeves and I explain in *Psychology through the Eyes of Faith* (2002), there are striking parallels between the human image in psychological science and in biblical and theological scholarship. Whether viewed through the lens of today's science or ancient biblical wisdom, human nature looks much the same. Some examples are as follows.

The unity of mind and body

- *Biblical and theological wisdom*: In Hebrew–Christian tradition, humans are embodied creatures, not immortal souls. We are bodies alive, and death – what St Paul called 'the last enemy' – is real. The afterlife is envisioned as a *new creation*, a resurrected *body*.
- *Psychological science*: In keeping with this tradition (but not with New Age dualism), today's cognitive neuroscience is ever tightening the links between mind and brain. Our minds do not operate without a brain. The very idea of thinking without a body is akin to running without legs.

Pride

- *Biblical and theological wisdom*: In the Christian tradition, pride is the fundamental sin – the deadliest of the 'seven deadly sins'.
- *Psychological science*: The well-documented counterpart to pride in today's psychological science is 'self-serving bias' – a powerful and often perilous tendency to perceive and present oneself as better than others.

Rationality and fallibility

- *Biblical and theological wisdom*: In the biblical perspective, humans are made in the divine image, yet they are finite and error-prone.
- *Psychological science*: In recent psychological science, the emerging image of humanness similarly notes both our remarkable cognitive capacities and our abundant illusory thinking (as the Nobel laureate Daniel Kahneman explains in his magnum opus, *Thinking Fast and Slow* (2011)).

Behaviour and belief

- *Biblical and theological wisdom*: Christian thinkers have often reminded us that faith predisposes action, yet it also grows through obedient action.
- *Psychological science*: 'Amen,' say social psychologists. Attitudes *influence* behaviour, and attitudes *follow* behaviour (as illustrated by racial attitudes changing *after* changed interracial behaviour, and by experiments in which people come to believe in their induced actions).

Religious engagement and human flourishing

Medicine abused can kill people. Medicine wisely practised enhances life. Is the same true of religion?

Religion abused kills. The insane courage that enabled the terror of September 11, 2001 'came from religion', noted Richard Dawkins (2001). But so has the motivation behind the founding of hospitals, hospices, universities and civil rights movements. Understandably, the evolutionist Stephen Jay Gould noted that much of his 'fascination' with religion lay 'in the stunning historical paradox that organized religion has fostered, throughout western history, both the most unspeakable horrors and the most heartrending examples of human goodness' (Gould, 1997).

While acknowledging religion's historic horrors and heroes, social scientists have explored religion's links with volunteerism, non-materialistic values and charitable giving. In survey after survey, people who are religiously engaged, or who say that religion is *important in their daily life*, exhibit, on average, greater generosity with their time and money. In Gallup World Poll data reported by Brett Pelham and Steve Crabtree (2008), for example, religiously engaged people in Europe, the Americas, Africa and Asia are about 50 per cent more likely to recall having donated to a charity in the last month (and this was despite religious people having *lower* incomes) (see Fig. 9.1).

Highly religious people were similarly more likely to report volunteering (see Fig. 9.2).

Is religious engagement likewise predictive of human happiness? The answer differs dramatically by whether we compare *places*

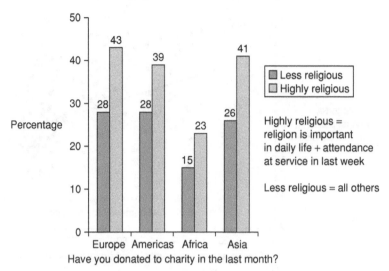

Figure 9.1 Religiosity and charity across the world
Gallup World Poll data reported by Pelham and Crabtree, 2008

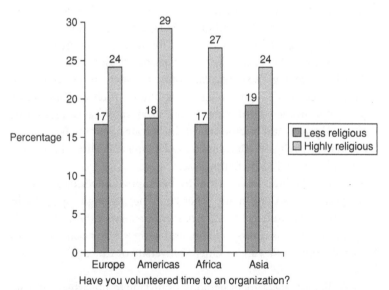

Figure 9.2 Religiosity and volunteerism across the world
Gallup World Poll data reported by Pelham and Crabtree, 2008

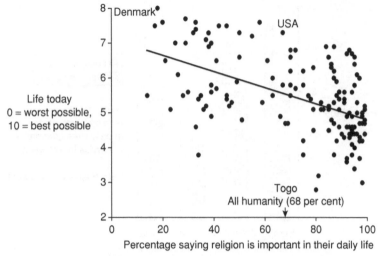

Figure 9.3 Religiosity and well-being across 152 countries
Gallup World Poll data harvested by the author

(such as more versus less religious countries or states) or *individuals*. Consider evidence for this 'religious engagement paradox'.

Harvesting Gallup World Poll data, I found a striking *negative* correlation across 152 countries between national religiosity and national well-being (see Fig. 9.3). Secular countries such as Denmark are happier places than highly religious countries such as Pakistan or Nigeria. Within the United States, I have also found that secular states, such as Oregon and Vermont, exhibit greater human flourishing than do highly religious states such as Alabama and Mississippi. Ergo, secular places tend to be happy places.

Yet religious individuals tend to be happy individuals. Survey data from the USA and many other countries (though especially in more religious countries) reveal a *positive* correlation between religiosity and happiness *across individuals*. Figure 9.4, for example, gives National Opinion Research Center (University of Chicago) data from the USA.

The religious engagement paradox – differing results when comparing *places* and *individuals* – appears in many other ways as well. In the *less* religious US *states*, people live longer, smoke less, commit less crime, have lower teen pregnancy rates – and the list goes on. Nevertheless, *more* religious *individuals* live longer, smoke less, commit less

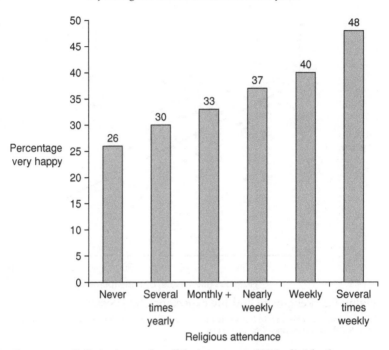

Figure 9.4 Religiosity and well-being across US individuals

National Opinion Research Center General Social Survey data harvested by the author

crime, have lower teen pregnancy rates – and the list, again, goes on. An example in Figure 9.5 overleaf shows that people in more religious states have *shorter* life expectancy.

Yet actively religious individuals have *longer* life expectancy (even after controlling for sex, age, race and region), as indicated in Figure 9.6 overleaf.

The Princeton economist Angus Deaton and the psychologist Arthur Stone have also been struck by this religious engagement paradox. They ask, 'Why might there be this sharp contradiction between religious people being happy and healthy, and religious places being anything but?'

Similar paradoxes (from aggregate versus individual data) occur in other realms:

- *Politics*: In the USA, low-income *states* tend to favour Republican presidential candidates while low-income *individuals* tend to favour Democratic presidential candidates.

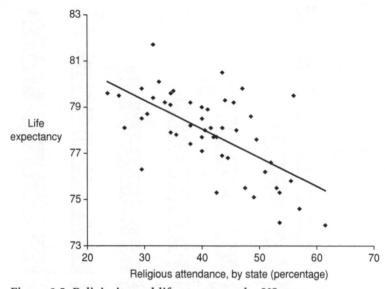

Figure 9.5 Religiosity and life expectancy by US state

Life expectancy data from Social Science Research Council's American Human Development Report 2008–9. Religious attendance data from Gallup

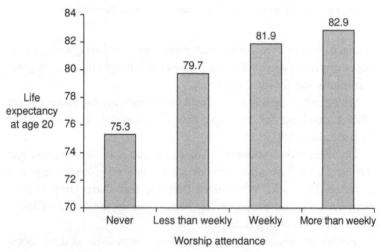

Figure 9.6 Religious attendance and life expectancy

National Health Interview Survey from Hummer, R. A., Rogers, R. G., Nam, C. B. and Ellison, C. G. (1999), 'Religious Involvement and US Adult Mortality, *Demography* 36, pp. 273–85

- *Conservatism and subjective well-being*: In Europe, more liberal *countries* and more conservative *individuals* express greater subjective well-being.
- *Pornography use*: People in highly religious *states* do more Google searching for sexually explicit content such as 'gay sex'. But highly religious *individuals* do less such pornography searching.

These are the sorts of findings that excite behavioural science sleuths. Surely there must be some confounding variables. With religiosity, one such variable is income – which is lower in highly religious countries and states. If researchers control for status factors such as income (as Ed Diener, Louis Tay and I have done), the negative correlation between religiosity and well-being disappears and even reverses to a slightly positive correlation. Likewise, low-income states differ from high-income states in many ways.

So, in matters of religion and politics, should we draw our conclusions more from aggregate or individual-level data? If you want to make religious engagement look bad – confirming Richard Dawkins' surmise that 'faith is one of the world's great evils' (Dawkins, 1997) – use the aggregate, macro-level data. If you want to make religious engagement look good – confirming that a 'fruit of the Spirit' is 'joy' – use the individual data. That acknowledged, is not the more important story found where life is lived – at the level of the individual, where *religious engagement predicts human flourishing* (and where high income predicts conservative voting)?

Prayer experiments

Amid these striking parallels between big biblical and psychological ideas and the evidence of the benefits of religious engagement, there have been two points of tension. One concerns prayer. Some studies identify thinking errors (such as *illusory correlation* and *the illusion of control*) that underlie superstitious thinking in realms such as gambling, stock investing and beliefs about supposed extrasensory perceptions. These tendencies to believe that one thing causes another when they really are only coincidentally correlated, and to assume that we are controlling events that are actually beyond our control, could easily lead people to perceive their prayers as effective, whether they are or not.

So are intercessory prayers effective? Is prayed-for rain more likely to fall on parched earth? Are people more likely to sail through cardiac bypass surgery if many people are praying for them (i.e. is prayer a medical antidote?)? As I explain in *A Friendly Letter to Skeptics and Atheists* (Myers, 2008), a series of actual experiments that tested a magical understanding of prayer consistently indicated 'no' (as I had publicly predicted).

If it is heretical to think too little of the power of our prayers, is it not equally heretical to think of God as a celestial Santa Claus? My conclusion, in anticipation of the medical prayer tests:

> Do we err in searching for a 'God effect' that is a slight subtraction to, for example, the number of stillbirths or coronary deaths? In the historical Christian understanding, God is not a distant genie who we call forth with our prayers, but rather the creator and sustainer of all that is. Thus, when the Pharisees pressed Jesus for some criteria by which they could validate the kingdom of God, Jesus answered, 'The kingdom of God is not coming with things that can be observed. . . . For, in fact, the kingdom of God is among you' [Luke 17.20–21, NRSV].
>
> The Lord's Prayer, the model prayer for Christians that I pray daily, does not attempt to control a God who withholds care unless cajoled. Rather, by affirming God's nature and our human dependence even for daily bread, it prepares us to receive what God is already providing. One can approach God as a small child might talk with a benevolent parent who knows the child's needs but also cherishes the relationship. Through prayer, people of faith express their praise and gratitude, confess their wrongdoing, voice their heart's concerns and desires, open themselves to the Spirit, and seek the peace and grace to live as God's own people. (Myers, 2008, pp. 42–3)

Sexual orientation

No issue divides Christians more than their differing understandings of sexual orientation, and their attitudes regarding the marriage and ordination of those with same-sex attractions and a gay or lesbian identity.

In *What God Has Joined Together: The Christian Case for Gay Marriage* (Myers and Dawson Scanzoni, 2005), we sought to bridge the divide between traditionalists (who want to support and renew marriage) and progressives (who believe that sexual orientation is not a choice and is best lived out within the context of a committed partnership). My bottom line, as a marriage-supporting social scientist, is simply that (a) sexual orientation is a natural, enduring disposition, and (b) the world would be a happier and healthier place if love, sex and marriage routinely went together.

To expand that nutshell synopsis just a bit, psychological science now has substantial evidence indicating the following:

- All humans have a deep *need to belong* – to connect with others in close, intimate, enduring relationships.
- As one important example of such relationships, marriage contributes to flourishing lives – to healthier and happier adults, and to children who thrive when co-parented by two parents who love each other and together love their children.
- Toxic forces, especially radical individualism and the media modelling of impulsive sexuality, are corroding marriage and the health of communities.
- Sexual orientation is a natural (largely biologically influenced) disposition, most clearly so for men. Scientists have discovered a host of gay–straight differences, including differing brain centres, fingerprint patterns and prenatal influences.
- Sexual orientation is also an enduring disposition, which is seldom reversed by willpower, reparative therapy or an 'ex-gay' ministry.

But are there not postnatal environmental influences on sexual orientation? Perhaps a domineering mother or distant father (as some Freud-influenced 'reparative therapists' have believed)? If that were so then shouldn't boys growing up in father-absent homes (or in modern times with more absentee dads) more often be gay? But such appears not to be the case.

The biological reality of sexual orientation is also suggested by observations of same-sex sexual behaviours in several hundred species, including the 8 per cent of rams that shun ewes and instead seek to mount other rams (and whose brains display neural structures akin to those in gay versus straight human brains). And it is suggested by twin and family studies that reveal some genetic influence on sexual orientation – an influence confirmed by a recent genomic analysis of 409 pairs of gay brothers, showing genetic differences in parts of two chromosomes, one maternally transmitted.

You may wonder: given that same-sex couples do not naturally reproduce, why would *gay genes* (probably many genes having small effects) exist in the human gene pool? Consider a possible answer: several studies have shown that:

1 homosexual men tend to have more homosexual relatives on their mother's side;
2 the heterosexual maternal relatives of homosexual men tend to produce more offspring than do the maternal relatives of heterosexual men.

This suggests a *fertile females theory*: the genes that dispose women to be strongly attracted (or attractive) to men – and to have more children – may also dispose some men to be attracted to men. Thus, there may be reproductive wisdom to genes that, as a by-product, dispose some men to love other men. If so, is sexual diversity a part of the biological wisdom of God's creation?

But 'What about the Bible?' Out of 31,103 Bible verses, only seven speak directly of same-sex behaviour – and often in the context of idolatry, promiscuity, adultery, child exploitation or violence. Biblical scholars have been debating the import of these verses, and of the wider biblical moral wisdom.

Arguing for the traditionalist position (sex and marriage are only for heterosexuals) are books by Gagnon (2001), DeYoung (2015) and Jones and Yarhouse (2001), who include biblical wisdom in their discussion.

On the other hand, a revisionist evangelical perspective argues that the Bible supports a consistent sexual ethic for gay and straight people (Rogers, 2006; Brownson, 2013; Johnson, 2006; Achtemeier, 2014; Wilson, 2014; Gushee et al., 2014). These latter scholars argue that the Bible has nothing to say about an enduring sexual orientation (a modern concept) or about loving, long-term same-sex partnerships. Some also offer a Christian case for gay marriage, based on the human need to belong, on the biblical mandate for justice for everyone, and on the benefits of a culture-wide, marriage-supporting monogamy norm.

For those who embrace a 'Reformed and ever-reforming' faith tradition – one that esteems a spirit of humility and is open to the continuing work of the Spirit – a fresh look at Scripture is always welcome. Across time people of faith have repeatedly revisited Scripture and changed their minds about marriage:

- from welcoming arranged marriages to favouring romantic choice;
- from assuming polygyny to mandating monogamy;
- from viewing marriage as inferior to celibacy to seeing it as an equal calling;
- from assuming male headship to preferring mutuality;
- from shunning interracial marriage to welcoming it;
- from disciplining divorced people to welcoming them into our faith communities.

In each case, our religious ancestors found proof texts to support their cultural assumptions, and later biblical scholarship led us to read and respect Scripture in a fresh way.

Christians will surely agree that their sexual ethics should not simply follow cultural trends. Rather than seeking to discern God's wisdom by putting our finger to the wind, we should put our nose to the Bible. But we may, as a practical matter, also want to be aware of how the Church's perceived stance affects the winsomeness of its good-news message.

The conservative religious position against same-sex partnerships is having an apparent counter-evangelism effect, suggest the Harvard researcher Robert Putnam and Notre Dame sociologist David Campbell. They have noted (from US data) that 'The association between religion and politics (and especially religion's intolerance of homosexuality)' is 'the single strongest factor' in alienating young people from the Church. A recent Ford Foundation-funded US national survey (Cox and Jones, 2014) for the Public Religion Research Institute confirmed their conclusion:

> Among millennials who no longer identify with their childhood religion, nearly one-third say that negative teachings about, or treatment of, gay and lesbian people was either a somewhat important (17 per cent) or very important (14 per cent) factor in their disaffiliation from religion.

Attitudes about sexual orientation are rapidly becoming more accepting of gay rights and relationships. Moreover, there is a large generation gap, with most older adults opposing gay marriage and most younger adults supporting it. Given that the forces driving the attitude changes are likely to continue, and given generational succession, it appears that the culture war over gay marriage and gay ordination will gradually be resolved in the years to come, much as were previous culture wars over minority and women's basic rights. As this happens, perhaps the winsomeness of Christian faith can be renewed for younger adults.

Synopsis

- Faith-rooted humility mandates the ever-reforming empirical spirit which helped give birth to modern science and which survives in our efforts to love God with our minds by exploring the human creation.
- Psychological science and religious faith have many points of contact, as psychologists reflect on their underlying assumptions and values, apply psychological findings to the faith community, study the psychology of religion, connect their respective wisdom about human nature, study the associations of religious engagement with human flourishing, and explore points of possible tension between psychological science and personal faith.
- Psychological and biblical understandings of human nature are strikingly congenial. Both affirm a unity of body and mind, the powers and perils of pride, the capacities and limits of human thinking, and the interplay of belief and behaviour.
- Research indicates positive associations (across individuals) between religious engagement and human flourishing, as indicated by generosity, longevity and happiness.
- Studies of illusory thinking and intercessory prayer, and of sexual orientation, challenge the Church to affirm and practise its ever-reforming heritage in a spirit of humility and love.

10

Being a person: Towards an integration of neuroscientific and Christian perspectives

JOHN WYATT

What does it mean to be a person?

Is it possible to exist as a living human being but not to qualify as being a person in some sense? On the other hand is it possible to be a person but not a living human being? Can a chimpanzee or an advanced artificial intelligence be thought of as a person? Is the internal sense that we have of being a unique individual, having a 'first-person perspective', merely an illusion created by the neuronal machinery of the brain?

These may sound like purely abstract speculations of the type that philosophers and theologians are paid to address. But in reality our understanding of what it means to be a person has profound real-world consequences. Disagreements about 'personhood' are at the root of many recent debates, in fields as diverse as medical ethics, law, psychology, social sciences, and even artificial intelligence and robotics.

My professional work as a neonatologist, a medical specialist in the intensive care of newborn infants, was predicated on a particular understanding of human personhood. Many of my patients were extremely premature infants, some born at the limits of viability at 22 or 23 weeks of gestation and weighing 500 grams or less. Our ethical framework was to treat each baby as a unique and precious individual, to make decisions about intensive medical treatment which were in their own best interests and, wherever possible, to maximize their chances of healthy survival.

Personhood based on self-awareness

But not everybody celebrated and supported our activities. Some philosophers and ethicists have challenged the view that all newborn babies can be regarded as persons who are full members of the human community. For Peter Singer, a 'person' is a being who has a capacity for enjoyable experiences, for interacting with others and for having preferences about continued life. It is clear that a newborn baby is not capable of interacting in any meaningful way and is unable to have preferences about his or her continued life. Kuhse and Singer (1985, p. 133) put it like this: 'When I think of myself as the person I now am, I realize that I did not come into existence until some time after birth.' Hence a newborn baby is a human being but not a 'person'.

The philosopher John Harris has a definition that is similar although not identical. A person is 'a being who is capable of valuing their own existence . . . The value of my life is precisely the value I give to my own life' (Harris, 1995, p. 9). In contrast, Michael Tooley has a definition that highlights the sense of having a narrative extending over time: 'A person is a being who is capable of understanding that they have a *continuing self*' (Tooley, 1983, p. 28).

These definitions all point to the centrality of some form of self-awareness, and it is obvious that for all of them a newborn baby will not qualify as a person, whatever the precise definition. This has profound implications for medical decisions about sustaining or ending life. As Kuhse and Singer put it,

> Only a person can want to go on living, or have plans for the future, because only a person can understand the possibility of a future existence for herself or himself. This means that to end the lives of people against their will is different from ending the lives of beings who are not people . . . killing a person against his or her will is a much more serious wrong than killing a being who is not a person.
>
> (Kuhse and Singer, 1985, p. 134)

Singer goes further and argues that 'the decision to kill a newborn infant is no more – and no less – the prevention of the existence of an additional person than is a decision not to reproduce'.

The same kind of thinking about personhood leads to the conclusion that individuals with severe learning disorders, brain injury or advanced dementia also cannot be regarded as persons. Personhood becomes defined by high-level cognitive functioning, an advanced level of integrated brain activity. In fact, in order to be regarded as a person, you must have a completely developed and normally functioning cerebral cortex.

Those who meet the criteria of being 'persons' have moral rights and privileges. They deserve to be protected from those who would injure or kill them. They should be allowed to exercise their own choices or autonomy as much as possible. But the same rights and privileges do not extend to 'non-persons'.

Of course there are major problems with this kind of definition of personhood. Why should high-level cognitive functioning be the vital criterion that distinguishes those beings whose lives are precious and to be protected from those who are effectively disposable? Why should the functioning of the cortex, a layer of neurones millimetres in thickness, be the central and crucial moral discriminating feature between beings? It isn't obvious that my cortical functioning should be the defining feature of my human worth and significance.

If my personhood depends from moment to moment on the precise functioning of my cerebral cortex, then it becomes a remarkably fragile and contingent phenomenon. At the moment, as I write these words, my cortex is functioning reasonably well (at least I hope so!), and I can be regarded as a person. But if, later today, as I cycle to work, I am involved in an unfortunate collision leading to severe head injury and cortical damage, then I will no longer be a person. Of course if, following rehabilitation, my cortical function recovers to a sufficient level, then I will become a person again.

Can something that seems so fundamental to my human identity and significance be so fragile and vulnerable to the contingencies of everyday life? Suppose I suffer severe brain injury but have the prospect of gradual recovery to normal cortical function. Am I a person in the intervening period? If someone kills me during the recovery period, is he or she guilty of the serious crime of killing a person or the less serious crime of killing a non-person?

This is an extreme example of the questions and challenges that surround the sense we all have of being a 'continuing self', of having an identity that is preserved over time and space. Since all the molecules of my body and my brain are in a constant state of turnover with the environment, to what extent am I the same person as I was ten years ago?

At the heart of the philosophical perspective put forward by Kuhse and Singer and by Harris, among others, is the idea that you earn the right to be called a person by what you can do, by demonstrating that your brain is functioning adequately, by thinking and choosing. This taps into the modern liberal emphasis on personal autonomy. To be a person is to be autonomous – or self-governing.

Substance dualism

It is interesting that liberal political philosophy is profoundly *dualistic*. 'I', the mysterious inner self, must be free to choose and to determine what happens to me, including what happens to my body. The conscious self is somehow disconnected from the body and is seen as its controller, governor and master. This conception of the human individual is rooted in the philosophical perspective of *mind–body dualism* pioneered by Descartes. It was re-emphasized by Immanuel Kant, who stressed the significance and centrality of the autonomous thinking self.

The mind is conceived as a substance, a form of 'stuff', a thinking stuff that is different from the physical stuff of the body, and the two substances interact in a mysterious way within the brain. Although *substance dualism* was popular at the time of the Enlightenment, it has become deeply unfashionable within the modern neuroscientific community. The dominant position of modern neuroscience is that there is no mysterious thinking 'stuff' connected to the brain. Most neuroscientists are resolutely *materialist* or *physicalist* in their understanding. The brain is a physical, material organ like all the other organs of the body and hence consciousness and self-awareness must have a physical origin within the activity of brain cells.

Physicalist understandings of conscious awareness

Some have argued that our sense of being a conscious, unitary, choosing self is merely an illusion created continuously by our brains because it has some survival advantage for us as a species. Francis Crick called this the 'astonishing hypothesis': 'You, your joys and your sorrows, your memories and your ambitions, your sense of personal identity and free will, are in fact no more than the behaviour of a vast assembly of nerve cells and their associated molecules' (Crick, 1994, p. 3). In the words of the science journalist Matt Ridley, 'There is no *me* inside my brain; there is only an ever-changing set of brain states, a distillation of history, emotion, instinct, experience, and the influence of other people – not to mention chance' (Ridley, 2003, p. 278).

According to this perspective our conscious awareness, our 'first-person perspective', the internal sense of being a unitary, choosing self is merely an epiphenomenon. It's a kind of mental 'froth' that emanates from our neuronal machinery, but it has no causal significance for our bodily actions and behaviour. In a famous analogy, it is like the steam that comes from the funnel of a locomotive. It is created by the mechanisms going on within the cylinders, but it does not in itself influence the course of the engine. Our actions are determined by neuronal machinery, independent of our conscious thoughts and intentions.

There is a notable logical incoherence between the dominant liberal understanding of the autonomous choosing self that operates on the body, and the dominant physicalist understanding of the human brain, in which the sense of the unitary self is a pervasive illusion created by the working of the brain. As we saw earlier, within the influential thinking of Kuhse and Singer, Harris and Tooley, it is the autonomous choosing self whose preferences and choices must be respected. I, the self, can choose to do whatever I like with my own body. But this makes little rational sense within a physicalist understanding of brain function. There is no metaphysical distinction between the self and the body. The self is an emergent property of the body, derived and created by neuronal processes.

Libet's experiments and determinism

Some philosophers and neuroscientists have claimed support for a *deterministic* or *epiphenomenalist* perspective from a series of experiments first pioneered by the neuroscientist Benjamin Libet in 1983. Libet asked volunteers to press a lever at a time of their choosing, while recording the subject's electroencephalogram (EEG) continuously, as well as the precise time at which the person noted his or her awareness of an intention to press the lever. Libet found that EEG changes (called the *readiness potential*) indicating neuronal prepared-ness to muscular action were detected significantly *before* the time at which the volunteer reported an awareness of an intention to press the lever.

These experimental results, which have been reproduced in a range of subsequent studies, have been interpreted to support a form of *hard determinism*. It is claimed that the actions of the subject in pressing the lever are determined by unconscious neuronal mechanisms, and the subject's belief that his or her actions were a result of voluntary choice was in reality due to retrospective rationalization.

Experiments of this kind have been subject to a range of criticisms within the neuroscientific community, and their relevance to philo-sophical questions of free will and determinism is questionable. The experimental design is highly artificial and it is clearly far removed from normal motor behaviour. The experimental subject is primed to press the lever and is focusing intently on the single motor action that he or she has already determined in advance to perform. Motor function in human beings is not a good analogue of truly 'voluntary' behaviour. It is known to be largely mediated by subcortical processes that do not enter conscious awareness. For example, the complex control mechanisms underlying walking or driving a car operate largely at an unconscious level. Hence the precise timing of the sub-cortical and cortical processes underlying voluntary movement com-pared with the timing of the conscious awareness of the intention to move a finger do not seem strictly relevant to broad questions of free will. Certainly, it seems that Benjamin Libet himself did not interpret his experimental results as a demonstration of hard determinism.

An alternative philosophical perspective is that of *non-reductive physicalism*. Again, the brain is seen as entirely physical and material in nature, and there is no other non-physical or mental 'stuff'. How-ever, in this view it is possible for mental states to 'emerge' from physical neuronal processes in a way that leads to new possibilities, including 'top-down causation', in which mental states influence neuronal activity as well as the reverse. In other words, mental states can have causal efficacy in the physical world. How it might be possible for mental states to provide top-down causation within a purely physicalist understanding of the brain remains highly contested and controversial.

Physicalist perspectives and artificial intelligence (AI)

Contemporary philosophical debates about human brain function have taken on a new importance because of the rapid development of computer systems capable of simulating human-like intelligence. If our brains are merely 'computers made out of meat', as Marvin Minsky famously claimed, then there is no a priori reason why all aspects of human cognition and brain function cannot be accurately emulated within an advanced computer system. From a physicalist perspective, there is nothing within the human brain that cannot in principle be reproduced within an artificial system. And we would expect such an artificial system ultimately to display the same emergent properties as intentionality, agency, rationality and forethought.

This kind of reasoning lies behind the concerns expressed by a number of prominent scientists and technologists that advanced AI systems may represent a serious threat to the future of humanity. If all the evil intentions and the destructive actions of dictators and tyrants have emerged from the physical processes of human brains, why might they not similarly emerge from an advanced artificial mechanism? Why should we presume that the emergent intentions and rationalizations of advanced artificial intelligences will always be strictly benevolent towards the human race?

The human brain and the process of scientific research

Of course, all attempts to use our brains to understand how our brains work may be doomed to failure. It has been said that 'If the human brain was so simple that we could understand it, we would be so simple that we couldn't understand anything.' Nonetheless, the very process of undertaking scientific research into the nature of the brain makes stringent and unavoidable demands upon our brain activity.

In order to do scientific research, including neuroscientific research, you have to believe that it is possible for a human being using a human brain to investigate and determine the scientific processes and regularities on which the brain is based. If our mental processes were merely determined by unconscious neuronal mechanisms, it is hard to see how this should be possible.

As any experienced scientist will testify, planning and undertaking original research is an intensely creative activity. Using a previous example, Benjamin Libet and his colleagues created an original experimental paradigm. They chose (out of myriad possibilities) a single motor action which was highly stereotyped and which could be timed to millisecond accuracy. They then created a novel method by which the experimental human subject could time the onset of conscious awareness of an intention to act, using a rapidly sweeping pointer on a clock face. Yet, if hard determinism is correct, the sense the experimenters had of voluntarily choosing and creating their unique experimental paradigm was entirely illusory and the result of post hoc rationalization. The design of the experiment was determined by unconscious neuronal mechanisms. Similarly, their interpretation of their experimental results was in reality unconsciously determined.

The practice of scientific research depends on the belief that the researcher is *genuinely free* to create hypotheses and models, to design experiments, to assess evidence and to choose the most consistent interpretation of the data. But, if you are committed to a physicalist understanding of the brain, you have to ask whether these beliefs are logically coherent.

Darwinian orthodoxy teaches that the distinctive features of human brain function evolved because they provided a survival advantage on the African savannah. Since human conscious awareness and the ability for rational planning, forethought and creativity require complex and nutritionally expensive processes, they must have provided a significant survival advantage. But this would strongly suggest that the microstructure of our brains and all our conscious mental processes are orientated towards providing a survival advantage rather than determining the truth about objective reality. Why should we trust the rationalizations that appear within our conscious awareness?

Of course, the brain of the neuroscientist is not immune from these evolutionary processes and pressures. An entirely physicalist understanding of brain function leads one to conclude that all conscious beliefs, including the scientific conclusions of the neuroscientist, are likely to be unreliable. Our brains should be orientated towards survival, not truth, and whenever there is a conflict between the two, survival should win. Yet the extraordinary success of modern science indicates that some human mental processes are in fact highly adapted to analysing and comprehending objective reality, even when such reality is extremely abstract, complex and counter-intuitive.

Why is the universe comprehensible to human beings?

In 1916 Albert Einstein published his general theory of relativity, describing in precise mathematical detail how space and time are warped by the effects of gravity. His theory provided quantitative predictions on how a massive gravitational body would disturb space and time in its vicinity. On the basis of Einstein's equations, two phenomena were predicted: the 'geodetic effect' (warping of space–time vectors around a massive object) and 'frame dragging' (the amount a spinning object twists space and time with it as it rotates).

More than 80 years later, in 2004, NASA launched a satellite called Gravity Probe B to test the observed accuracy of predictions based on Einstein's equations. The satellite carried the most mechanically precise gyroscopes ever engineered in order to measure the amount that space–time is warped by the presence of the spinning Earth. It was calculated that the geodetic effect should cause the axes of the gyros to deviate by 0.0018 degrees per year, while frame dragging should cause a separate perpendicular movement of 0.000011 degrees per year. This was described as equivalent to detecting the thickness of a sheet of paper held edge-on 100 miles away. The conclusion of NASA's multimillion-dollar experiment was that the observations fit the warping of space–time predicted by Einstein's equations to the limits of experimental accuracy. The abstract equations worked in the real world to the most mind-boggling level of accuracy.

But why should complex and abstract mathematical equations, which emerged as a result of physical processes in the brain of a carbon-based life form, describe with astonishing accuracy the counter-intuitive behaviour of the Earth as it drags space and time around with it? As many philosophers have noted, there appears to be a strange connection, a homology, between the mind and thoughts of this pathetic and insignificant life form and the fundamental structures of the cosmos. As Einstein himself put it in his 1936 essay, 'Physics and Reality', 'The eternal mystery of the world is its *comprehensibility*.'

An alternative perspective: panpsychism

Reflections like this have led some secular philosophers to conclude that mental states and consciousness are not merely a coincidental product of neo-Darwinian survival mechanisms. Our mental states appear to have a mysterious representational connection with the rest of the cosmos. For example, the philosopher Thomas Nagel argues that the intelligibility of the world is no accident. He argues that minds are related to the natural order in two distinct ways. First, nature is such as to give rise to conscious beings with minds. Mind and conscious awareness emerge naturally and inevitably as a consequence of evolutionary mechanisms and increasing complexity. Second, nature is such as to be intelligible to conscious beings. The universe is more than merely physical; it is populated by minds of varying degrees of awareness and sophistication.

Nagel's proposal is a form of *panpsychism*, in which consciousness or mind is a universal feature of all physical objects, and the universe itself is orientated towards the emergence of advanced levels of conscious awareness and rationality. Nagel explicitly rejects a traditional form of Christian or theistic belief about a conscious and thinking creator, although he recognizes the explanatory power of the theistic worldview. Instead, he opts for a kind of natural *teleology*. The universe has the appearance of purpose, the fostering and emergence of mind, but it does not have a conscious creator or sustainer. Panpsychism is not a new idea, but for contemporary neuroscientists it is deeply unfashionable. Physicalists argue that it represents the worst kind of metaphysical speculation and that there is no empirical test that could decisively confirm or refute panpsychism.

The mutual dependence of science and faith in understanding personhood

The theme of this collection of essays is the mutual dependence of science and faith. In the area of personhood my contention is that science alone is not able to develop a full-orbed and comprehensive understanding. We depend upon a theistic and Christian perspective for a richer and more nuanced conception.

In the history of philosophy, the very idea of a 'person' has been strongly influenced by a Christian understanding of what it means to be a human being. The original Greek word for person (*prosopon*) meant literally 'the face', but in ancient Greek it also referred to the mask which actors used to represent the character they were playing in the theatre. In the world of Graeco-Roman thought what mattered about a human being was the face he or she showed to the world, the role he or she played in society. We have retained this meaning when we refer to someone's 'persona'. It is the public face that an individual shows to the world.

It is interesting that this is how the word is used in the Greek New Testament. At several points in Paul's epistles God is described as one who shows no favouritism. The literal Greek says that he is not 'a respecter of persons', meaning that he is not influenced by our external and social role.

However, in Hebrews 1.3 the Son is described as the exact representation of God's nature, and a different Greek word is used for the divine nature, the word *hypostasis*, which means literally 'what lies under'. The early Church Fathers, as they reflected on the nature of the triune Godhead, fastened on this word *hypostasis* as a means of describing the three individual actors within the Godhead. God's ultimate being (what 'lay under' his activity) was not merely substance, 'God-stuff', but was in the form of hypostases, persons giving themselves to one another in love.

Furthermore, the Christian revelation makes the remarkable claim that human beings are created in God's image. We are created to reflect the divine character and being. Because God's nature is personal, then we too are created and embodied as persons.

Persons are not reducible to matter and energy

Christian theism conceives of reality as consisting of more than just matter and energy. A purely materialist or physicalist description of reality will always be incomplete. There is another foundational category of reality and that is not 'mind' or 'consciousness' but the *personal*. As Martin Buber put it, reality consists not only of 'I–it' relationships but also 'I–thou' relationships. Persons are not reducible to matter and energy and they are not limited to matter and energy. In technical terms personhood is a category of reality that is ontologically foundational: persons cannot be defined in terms of other more basic categories such as 'substance' or 'rationality'. Persons are different from everything else in the cosmos.

Persons are *knowers* – they perceive and understand things about reality. Einstein came to a profound understanding of the nature of space and time expressed in his theories of relativity. Persons are *agents* – they do things, they have intentions and volitions, they make things happen. Experimental scientists design and carry out experimental procedures with the intention of testing or falsifying hypotheses.

Persons are *rational* – they understand and use logical analysis to comprehend and change reality. Persons are *communicative* – they speak with the intention of being understood by other persons, and the expectation that their communication will be successful. Persons are *creative* – they are genuinely innovative and free. Persons are *moral* – they understand the concepts of good and evil and they are accountable for their moral choices. Persons are *lovers* – they enter into profound and committed relationships with other persons.

None of these characteristics of human persons can be explained by or reduced to the nature of the physical universe, to the non-personal characteristics of matter and energy and physical laws. Persons are a different kind of reality.

In Christian thinking, because our human identity is derived from the being and person of God himself, human personhood cannot be self-explanatory. Philosophical reflection and neuroscientific analysis can never fully determine what it means to be human. The structure of our humanity, and the values and purposes of our human lives only make sense in the light of our creation in God's image.

A theistic understanding of humanity underpins the scientific enterprise

A theistic understanding of humanity also provides a theoretical framework in which the scientific enterprise can be placed. Because we are God-like beings, our thinking, our mental processes and subjective awareness are somehow homologous to the mind of God, and hence to the fundamental structures of the cosmos. So the Christian faith provides a conceptual framework, an epistemology, in which the homology between the mind of the human scientist and the structures of the universe makes sense. As the astronomer Johannes Kepler put it, 'I am thinking God's thoughts after him.' It is not surprising that many historians of science trace a causal link between theistic thinking and the rise of modern science in the seventeenth and eighteenth centuries.

Personhood in the light of Trinitarian thinking

Just as the three persons of the Trinity are individually unique, and yet give themselves continually in love, so each human person is unique, yet made for relationship with others. 'Personhood' is not something we can have in isolation; in Christian thinking it is a relational concept. Persons are constituted by their relationships – their very being is derived from communion and love, from the freedom to give oneself to the other.

Descartes' famous statement, '*Cogito ergo sum*, I think, therefore I am', places individual conscious awareness as the bedrock of existence. When everything else can be doubted, the final place for certainty is my own self-awareness. But this way of thinking leads inexorably to a profound individualism, which is one of the recurring features of modern life.

By contrast, we might suggest an alternative Christian version: '*Amor ergo sum*, I am being loved, therefore I am.' My being comes not from my rational abilities or self-awareness but from the fact that I am known and loved by others. In the words of the Christian philosopher Josef Pieper, 'Love is a way of saying to a person, "It's good that you exist; it's good that you are in the world"' (Pieper, 1997, p. 164).

In my professional work as a neonatologist, these Christian reflections led to a new recognition of the innate personhood of critically ill, preterm babies in my care. Instead of focusing on their limited functional abilities and reducing them to the status of a 'non-person', or merely a 'potential person', Christian thinking calls us to recognize the baby as a mysterious other, a 'thou' not an 'it', and a person to whom we as professionals owe a duty of care and protection. It is notable that contemporary understandings and practices of medicine and law in neonatology still reflect a Christian understanding of personhood from the moment of birth, rather than the preference utilitarianism of Singer and colleagues.

Conclusion: the unity of the human person

To conclude: it seems to me that both Christian thinking and contemporary neuroscience resist the substance dualism of Descartes, and emphasize the unity of our being. Human beings are not made out of two different substances. But Christian thinking cannot accept a physicalism that fails to give ontological respect to the immaterial aspects of being human. From a historic Christian perspective the human being is seen as a profound unity, a unity that has both a physical, material aspect and an immaterial, personal aspect.

How these two different aspects – the material and the immaterial – interrelate and integrate within the unity of the human person is deeply mysterious. Perhaps there is a parallel in the traditional theology of Christology. The Council of Chalcedon in AD 451 wrestled with competing understandings of the being of Christ. The Church Fathers were at pains to preserve the unity of the personhood of Christ despite the orthodox understanding that Christ was both human and divine. They eventually agreed on the formula that 'Christ is one person (*hypostasis*) who is one being (*homoousion*) with the Father, and one being (*homoousion*) with us – one person in two natures'. In the orthodox formulation Christ is both fully human and fully divine.

Of course it is unwise to press the comparison between our own human nature as created, limited and embodied beings, and that of Christ himself. But with due caution it may be possible to speculate that the unity of the human being – one person with both material and immaterial aspects – parallels in some mysterious way the profound unity of the Second Person of the Trinity. This 'Chalcedonian ontology' is neither monist nor dualist, neither physicalist nor idealist. The profound mystery of the human person transcends these distinctions. Each one of us is an integrated unity constituted by our relations with other persons, at the same time fully material and fully immaterial.

11

From projection to connection: Conversations between science, spirituality and health

JOHN SWINTON

The relationship between religion, spirituality, science and health has at times been rather troubled and fraught, particularly within the realm of healthcare. Freud's powerful proposition that religion is 'nothing but' a projection of human anxiety on to a transcendent screen, that it is a subtle form of neurosis that needs to be treated and eliminated, has formed a difficult backdrop for those working not only within the mental health professions, but within healthcare in general (Paley, 2008). Across the board there has been and continues to be a wariness towards *religion* – and its more recent companion *spirituality* – and an uncertainty as to just how necessary or benign it actually is. This difficult tension is exacerbated by the increasingly loud voice of the so-called 'New Atheists' and their neo-Epicurean rhetoric, combined with the worldwide rise of violent fundamentalist religion. It is therefore a challenging time to attempt to open up a positive conversation around the relationship between religion, spirituality, science and human well-being.

Nonetheless, despite these pressing issues, the conversation between religion, spirituality, science and health may in fact be vital insofar as it has the potential to bring together, in a positive way, perspectives on the world which in other areas have become separated and deeply fragmented. Perhaps the spiritual politics of the moment make such a conversation vital.

In this chapter I want to lay out some of the issues with regard to the ways in which science and religion can come together in the mutual quest for health and human well-being. Rather than being adversaries, properly understood they have the potential to provide complementary perspectives that can help us understand more fully the nature of human well-being and how that might be achieved and sustained. I will begin by exploring some of the scientific findings that indicate that religion and spirituality may have health benefits, before moving on to present an example of how spirituality and science can come together in ways that are credible and life enhancing. Along the way, I will offer some challenges to those who consider religion to be a dangerous socially constructed projection that has no credible place within the development of human well-being. As an alternative, I will suggest that such criticisms are in fact indicative of a deeper problem within Western societies rather than a movement towards 'enlightenment'.

Clarifying our terms

We need to begin by clarifying the terms of engagement. What exactly do we mean when we talk about religion and spirituality, and why would we separate these two terms? One of the fascinating things about postmodern spirituality is the way that religion and spirituality have come to be perceived as two separate concepts. It is important to begin by clarifying what each of these concepts means and how and why they have come to be separated.

Spirituality

It is frequently stated that *spirituality* is something that *all* people have and experience, but *religion* is assumed to be just one particular expression of this generalized universal spirituality. Within such a framework spirituality is assumed to be a kind of personal, existential quest for meaning, purpose, hope, value, love and for some people the divine and the sacred. Spirituality in this understanding is not confined to religious practices, rituals and communities. Rather, spirituality is expanded into a broad range of human experiences that are diverse and sensuous rather than tied to any single intellectual tradition. People may express their spirituality by engaging in yoga, mindfulness, the creative arts, coming close to nature by walking in the hills or gardening; they may work out their spirituality through their family, their community, work, or any other experience that draws them to a place which provides meaning, purpose and fulfilment to their lives (Heelas and Woodhead, 2005). Spirituality has come to be understood as an umbrella term that encompasses *all* people and relates specifically to the existential requirements of individuals. It is worth noting the close connection between culture and spirituality. This type of open, personalized, non-foundationalist spirituality fits neatly into the expectations and desires of contemporary Western women and men. In a very real sense, spirituality in this mode is a form of self-actualization. It is *my* personal meaning, *my* values, *my* purpose, and so forth. Spirituality is assumed to be an existential accompaniment to self-development. It is certainly the case that for some people altruism and community are central to that understanding of spirituality. But the desire to engage in such activities is perceived as a *choice*, rather than an obligation or a natural intuition.

Religion

Within this spiritual milieu, *religion* is conceived as a specific manifestation of a general phenomenon. All people are spiritual, but only some choose to be religious. While spirituality is presumed to be open and diverse and individual, the general assumption is that religion relates to more specific practices and beliefs and the more concrete shared community. Unlike spirituality, religion has roots, histories, traditions, and theological and philosophical underpinnings that are made explicit within doctrine and practice. Religions are mostly focused around particular gods or divine figures (with the exception of a religion like Theravada Buddhism where there is no omnipotent Creator God; gods do exist, but they exist as various spiritual beings and have limited powers). Religions are shaped and formed by specific traditions, and very often intentionally create particular forms of community. They contain narratives, rites, communal activities and worldviews that enable people to situate themselves in the world in quite particular ways. An important difference between spirituality and religion is the location of the spiritual impulse. The more generic forms of spirituality assume that spirituality comes from the inside; that is, it is an expression of human desire. However, for religious people spirituality is assumed to come from outside themselves, that is, it is given to them by way of an external force – God – in order that they can be enabled to live their lives together with that God and other followers in quite particular ways.

Measuring the spiritual

With regard to bringing science and spirituality together, one problem has been that the more generic spirituality is very difficult to tie down and to define. Because it is so personal and because it really cannot be generalized in a meaningful sense, it is very difficult to measure. If one reads through the various definitions of spirituality within the literature, it quickly becomes clear that the meaning of spirituality is determined by the predilections of the particular researcher writing the report. In other words, there is no generalized definition of the concept of spirituality that everyone will accept as a norm. Definition will therefore inevitably be vague, diffuse and thin. That is not, however, to say that the concept of spirituality is not important. Issues of meaning and purpose and value are fundamentally important for the way in which we structure and frame our lives and particularly how we structure and frame our experiences of illness and dis-ease. The more open generic perceptions of spirituality draw attention to this. In a biomedical context where technology and technique can overwhelm the personal, a focus on spirituality becomes particularly important even if it seems destined to remain ill-defined (Swinton and Pattison, 2010).

Generic spirituality may not be measurable using standard scientific tools, but it is certainly open to qualitative forms of research that seek to explore and understand the lived spiritual experience of people in a given situation. This is why the vast majority of scientific exploration that focuses on this broader model of spirituality is social scientific in origin and qualitative in outcome. This is not of course in itself a problem. Gaining knowledge and insight into the experience of spirituality as it relates to health, illness and well-being is fundamentally important for the practices of caring. Nevertheless, within a medical context that has inherent epistemological leanings towards the quantitative, creating positive relationships between the qualitative nature of generic spirituality and the quantitative demands of the hard sciences can be difficult to achieve and sustain. If there is no unity around what spirituality is, if people are unclear as to what its fixed unchanging constitutive elements are, then the scientific method will inevitably struggle to produce definite conclusions.

Measuring religion

Religion, on the other hand, does in some ways lend itself to scientific investigation. The reason for this is that aspects of religion are much more open to empirical observation and measurement. This is one of the reasons why the scientific method has become popular, particularly within United States, as a way of measuring the efficacy of religion in relation to the health and well-being of participants. Beginning in the early 1990s (Miller and Thoresen, 2003), researchers began to draw upon the methods of science in order to measure the relationships between religion and health in ways which are methodologically sophisticated and which seek to be credible not only to religious organizations, but also to the scientific community. Miller and Thoresen note that:

> Before the 1990s, the relationship between religion and health was largely a de facto area of research: researchers often buried religious variables in the methods and results sections of their studies without overtly highlighting them as legitimate areas of health research.
>
> (Miller and Thoresen, 2003)

After the 1990s there was a movement that sought to shift this area of research from the peripheral to the mainstream of scientific inquiry. With the publication of several special issues on spirituality and health in professionally refereed scientific journals (*American Psychologist*, 2003; Baumeister and Sedikides, 2002; Thoresen and Harris, 1999), what we might describe as the 'religion and health movement' (Levin, 2001) was well and truly under way. While there are a number of different perspectives on religion and health emerging from the USA, it is this scientifically orientated approach with its mantra of 'Religion is good for your health!' and its focus on religion in terms of structure and behaviour that has attracted a good deal of attention and the majority of the funding.

The way that religion is framed within this body of research takes the outward workings of religious traditions and practices and uses them as a gauge as to whether or not they are health bringing. So, for example, church attendance and involvement in religious communities have been shown to be protective against anxiety and depression (McCullough and Larson, 1999). Religious affiliation enhances social connection, which in turn is beneficial to both mental and physical health. Likewise, such religious practices as prayer, meditation and mindfulness have been shown to be beneficial for mental health; that is, bringing a sense of peace, well-being and physical health. This structural behaviourist perspective on religion presents evidence to suggest that some of the health benefits of religion would include: extended life expectancy, lower blood pressure, lower rates of death from coronary artery disease, reduction in myocardial infarction, increased success in heart transplants, reduced serum cholesterol levels and reduced levels of pain in cancer sufferers (Koenig et al., 2001). The reasons why this might be the case include:

- the social element of religious belonging – knowing people care for you, and having people do practical things to help you;
- the promotion of positive self-perception – 'God/the Divine cares for me; I am a child of God';
- the provision of specific coping resources – symbols, rituals and narratives that faith communities provide to give a framework for life;
- the generation of positive emotions – for example, love and forgiveness;
- encouraging people to be 'outward looking' – caring for others, the community and the environment helps people stay connected to the life around them and encourages a sense of purpose;
- the sense of meaning that religion can bring, providing answers to some of life's 'big' questions (Sims, 1994).

A key aspect of this approach is the way it seeks to utilize the methods of science to determine the relationships between religious practices and health. Standard scientific techniques such as randomized control trials, statistical analyses and modes of research that follow the principles of falsifiability, generalization and replicability mark this approach out as firmly within the paradigm of the so-called 'hard sciences'.

This approach is not without its theological and scientific critics (Schuman and Meador, 2004; Sloan, 2006). From the perspective of science and medicine, Richard Sloan has criticized the work done within this area as being methodologically questionable, based on the assumption that there is a unified, universal entity called 'religion'. Religion is a diverse and complex term and people adhere to religions in equally diverse and complex ways. To correctly and accurately identify the origins of the health benefits that people are perceived to have gained from participation in religion can be extremely difficult. The psychologist Kenneth Pargament has pointed out, in his work on religion and coping, that it is not so much the generalities of religion that bring about benefits, but the particularities of specific practices and specific beliefs (Pargament, 2002). In other words, it can be the *meaning* of a particular practice for the individual, rather than the religious tradition as a whole. The banner claim that 'religion is good for your health' requires a great deal more nuance than some of the studies allow.

Probably the most convincing research is that done on the health benefits of involvement in religious communities and in particular church communities. The evidence seems to indicate that adherents of religious communities have better health and are less likely to suffer from such things as depression and anxiety than those who do not belong to such communities. This is helpful, but it doesn't tell us as much as it claims. In line with Pargament's criticism, this doesn't, for example, tell us anything about *why* people are attending the churches or precisely *what* it is that they are getting from them. As Richard Sloan has put it: 'Anyone who believes that sitting in church makes you a Christian must also believe that sitting in a garage makes you a car!' (Sloan, 2006, p. 151).

Scientifically, this way of researching religion is problematic. It may well tell us some things about the way in which religious practices enhance human health and well-being. But, in order to do that, it has to reduce religious practices to behaviours. So, we can use science to observe and measure the changes that occur during prayer, meditation or worship; that is, as they are manifested purely in a material sense. The fact that one's blood pressure reduces when one meditates is all that science can tell us. The suggestion that the meditation has intentionality, that it is *about* something greater than the physical act, cannot be observed because consciousness cannot be observed. Materiality alone cannot have intentionality; matter is not *about* anything. Likewise prayer may be observed and measured in relation to the way in which it makes an impact upon the immunological system. The fact that the healing movement may be initiated by a divine presence simply cannot be part of the equation. Scientific enquiry into religion can tell us some things, and these things are important. But it cannot tell us everything that we need to know about religious experience.

Theologically, this way of looking at the relationship between religion and health is problematic. With the possible exceptions of the Christian Scientist movement and the Seventh-day Adventists, it is arguable whether the intention of most religious traditions is to focus on the attainment of personal health, at least not health as it is defined by the medical model (health equates to the absence of illness, anxiety or distress). Indeed, many religions call for sacrifices which can be significantly detrimental to health. It would be an interesting exercise, for example, to go through the Baptist missionary records and look at the health experiences of the early missionaries. I would suspect that religion was often very *bad* for their health! Within the Judaeo-Christian tradition, health is defined not by the *absence* of illness, pain, suffering or anxiety, but by the *presence* of God. The Bible has no word for health as we might understand it within biomedical culture. The closest biblical concept that we have to an understanding of health as we might think of it today is the Hebrew term *shalom*. Shalom relates to righteousness, holiness; being in right relationship with God. Within this understanding, to be healthy is to be in a right relationship with God (Wilkinson, 1980). To be in right relationship with God is dependent not on the absence of distress, but on the presence and the faithfulness of God. The intention of most religions is to enable people to relate to God, and this is a process that is not determined by the health effects of particular sets of beliefs and practices. The danger with using science to measure the health benefits brought about by religion is that, in looking at the outward manifestations of religion, it completely overlooks the inner processes of what it means to be a religious person.

Science then can certainly tell us some important things with regard to how religions function at a structural behavioural level in the process of facilitating human health and well-being. However, another dimension is required if the fullness of that experience is to be captured and understood. This requires a level of hospitality between science, religion and spirituality. We will return to the nature of such hospitality towards the end of this chapter. For now it will suffice to note that, in spite of these challenges and critiques, scientific exploration of religion that approaches its task respectfully and hospitably can serve to begin to break down some of the ingrained prejudices against religion that are so prevalent within the realms of culture and healthcare. In a secular and secularizing culture that can be deeply hostile to issues of religion, being able to evidence particular instances wherein religion can be shown to function positively and healthily can serve as a creative and constructive aspect in the ongoing and difficult conversation.

Spirituality and science: the biology of God

I previously indicated that the more generic understanding of spirituality poses some challenges with regard to whether or not the hard sciences are able to analyse it in any kind of meaningful way. I suggested that qualitative approaches were probably the best mode of approach to this kind of spirituality. That is not to suggest that the hard sciences have nothing to contribute to the conversation. A good example of one way in which religion, spirituality and science can be brought together is found in the work of my former colleague at the University of Aberdeen, Dr David Hay, and his research on what he describes as 'the biology of the spirit'. His basic hypothesis is the suggestion that human beings are innately spiritual; that is, that they are in some way 'hardwired for spiritual experience'. The diversity within spiritual experiences can be understood and explained using social-scientific methods alongside data from the hard sciences.

Spirituality as 'natural'

David Hay was a zoologist and a researcher at the University of Aberdeen's Centre for Spirituality, Health and Disability (CSHAD). Hay's central interest is in human spirituality. The discipline within which he began his research is zoology. He builds his basic hypothesis on the thinking of fellow zoologist Alister Hardy and his proposition that human spirituality has biological roots. In his 1966 Gifford Lectures at the University of Aberdeen, Hardy put forward the hypothesis that religious and spiritual awareness have evolved in the human species because they have proved to have survival value (Hardy, 1965, p. 292). Over many years, David Hay developed Hardy's work and put together some convincing scientific evidence indicating that spirituality and spiritual experience have a biological basis; that is, that human beings are in some sense 'hardwired' for spiritual experience. In his extensive qualitative research into the spirituality of children laid out in his book *The Spirit of the Child* (Hay and Nye, 2006), Hay argues that human beings have an inherent natural spirituality, which is manifested clearly in the lives of children. This natural spirituality reveals itself in the way that children have an intrinsic sense of awe, wonder, relationality and connectedness, as well as an acceptance of things beyond their understanding. While Western people have a tendency to assume that they are free, autonomous individuals, the natural relationality of children, which Hay calls 'relational con-sciousness' and equates with spirituality, indicates that human beings are by nature interconnected and relational. The gaining of autonomy requires not freedom from the other, but interdependence with others. We become autonomous not by being free from the other, but by being in relationship with others.

Among other things, Hay argues that the way we educate children suppresses and represses their inherent spirituality. When children are taught to prioritize rationality, reasonability, indi-viduality and 'hard facts' over feelings, their inherent relationality becomes repressed and de-prioritized. Hay suggests that this con-tributes to some of the problems that teenagers encounter when they are forced to become individuals while at the same time being driven by an innate relationality. The cultural emphasis on rationality and independence disguises the fact that the primal sense of awe, wonder, relatedness and connectedness that we find in children is in fact the 'true state' of all human beings: *we are relational creatures.*

Spirituality as relational consciousness

Hay describes this inherent relationality that is obvious in children but often hidden or emergent in adults as *relational consciousness*. Relational consciousness is a form of consciousness characterized by the fact that it is always relational: self–other people, self–environment, self–God. It is what makes spirituality possible and in a certain sense 'is' spirituality. Phenomenologically, it is experienced as the shorten-ing of the psychological distance between self and the rest of reality; a dissolving of the boundaries, which at the limit becomes the loss of distinction between self and other of the mystic. Relational consciousness occurs or rather is experienced when one realizes one's interconnectivity with others, with God (for some) and with the world in which one resides. As Hay puts it:

> A new awareness of the singularity of reality brings with it the realization
> that I am much closer to other people, the environment and God than
> I had originally thought. In relation to other people and the environment,
> I discover in myself an obligation to care for them, even to the point
> of self-sacrifice . . .   (David Hay, personal correspondence)

Relational consciousness is thus seen to be a natural urge or desire
to reach out and connect with someone or something beyond one's
self. However, and importantly, such spirituality is accompanied
by a sense of having an obligation to care for other people, even to
the point of self-sacrifice (Hay, 2011). Relational connectedness and
altruism form the essence of relational consciousness in Hay's understanding of spirituality. The point of spirituality then is connection,
not necessarily well-being. In other words, spirituality is not intended
for *health* (although that may be a by-product); rather it is intended
for connection. From Hay's perspective this primal experience is
the source of the experiential basis of religion, seen as a spiritual
construction in response to this natural spiritual experience.

In supporting his hypothesis, Hay draws from a wide range of
empirical scientific evidence. For example, he uses twin studies
designed to distinguish human characteristics that are inherited
from those that are acquired from the environment. Likewise, he
delves into genetics, exploring the evidence for the suggestion that
spiritual awareness is significantly linked to genetic inheritance
across cultures (Kirk et al., 1999). He also draws on studies from
neurology which indicate that there seems to be a biological basis
for the experience of religious and spiritual experiences (Newberg
and D'Aquili, 2001). It is from within this hospitable conversation
between qualitative and quantitative research that he claims to have
made a profound discovery about the origins and significance of
human spirituality.

One objection to Hay's hypothesis might be that, within it,
spirituality does not necessarily require God. If it is 'nothing but' a
biological function, then does that not simply reduce spirituality to
the neurological and meaningless consequence of blind evolution?
Hay would argue that this is not necessarily the case. His response
would be something along the following lines. Evolution always occurs
in response to something that happens in the world. So an eye develops
over time because of things to see. An ear develops over time because
of things to hear. Hay would argue that a sense of the spiritual, the

idea of relational consciousness, develops over time because the spiritual is something to be conscious of. So from his point of view there is no clash between theories of evolution and the idea of spirituality as relational consciousness.

If we accept Hay's hypotheses, two important consequences emerge. First, human awareness of the spiritual is seen to have a physiological component through which the individual comes to feel that he or she is an intimately connected part of a whole. In highly secularized cultures, this spiritual insight may not reveal itself in the form of religion, but may be manifested in a broad range of relational contexts: partners, family, community, God, religion and so forth. Hay's theory therefore contributes significantly to our understanding of why it is that, despite the continuing decline of traditional religions, a high percentage of people still claim to be searching after the spiritual.

Second, if human beings are by nature relational, and if spirituality is in fact a human universal, this provides some fascinating insights into the nature of contemporary Western cultures and a powerful counter to some of the more strident critics of religion. Both Sigmund Freud and Karl Marx assumed that religion and spirituality were nothing more than social and psychological constructions, designed to alleviate anxiety or to support the political power of the upper classes. For both, in slightly different ways, the well-being of people was assured when religion was seen for what it is: a human projection, a form of neurosis or an oppressive tool of a powerful elite. Religion understood in this way is nothing but a social construct. What is interesting about Hay's proposition is the way in which he uses science and the theory of evolution to make the radical counterpoint that it is in fact contemporary Western individualism and secularism that are socially constructed. If human beings are biologically wired for spiritual–relational encounter, then it is individualism and secularism that are pathological social constructions. Put slightly differently, the waning interest in religion and spirituality within Western cultures (if in fact there is a waning interest, which in Hay's perspective is open to debate) is in fact a significant cultural, psychological and spiritual problem that is indicative of the ways in which our natural biological propensities towards spirituality are being usurped and suppressed by a cultural tendency to understand the self in isolation rather than in relationship. Rising rates of loneliness, alienation, suicide, depression and anxiety would appear to bear witness to some implications of what Hay is proposing. The movement towards secularism, far from being a form of enlightenment, is simply another indication of the ways in which Western culture's pathological tendency to prioritize choice and individualism over commitment, altruism and relationality is working itself out within the lives of society's most vulnerable people. Individualism and separation from others (including God) is our *perceived* state, a constructed perception of the world within which we think we live. Relational consciousness reveals our natural state as beings-in-relationship.

Hay's perspective brings together science, religion and spirituality in a way that enlightens something fundamental about what it means to be a human being. In so doing, he provides deep insights into why, even within a radically secularizing Western European context, people still seek after the spiritual.

Hospitable connections

It seems fairly clear that the connection between religion and science within the area of health and human well-being does not have to be viewed as inevitably negative. As we have seen, the hard sciences do indeed have something important to offer, as do the social sciences (although the idea of hard and soft science is clearly challenged by Hay's perspective). It is certainly the case that, on their own, the sciences in all their different forms cannot and should not claim to provide all that we need to know about how religion, spirituality and health interrelate, but their contribution remains potentially significant. We can now begin to evidence ways in which religion can function creatively and positively in the process of facilitating healing, stabilizing people in the midst of demoralizing illness, and creating opportunities for innovative ways of facilitating coping. This is helpful and indicative of a way in which scientific study can help clarify misunderstanding and misapprehensions about the role of religion and spirituality. It is certainly the case that there are significant methodological and theological issues that need to be borne in mind, but such concerns only point to the need for a creatively critical approach with regard to how we frame and understand the relationship between these different forms of knowledge. Likewise the type of integrating work carried out by people like David Hay indicates ways in which qualitative and quantitative research can not only help us to understand what spirituality is, but also serve to counter significant misunderstandings about the nature and form of the cultures that we create and inhabit.

There are of course dangers that need to be monitored and foregrounded. Religion, like any powerful body of knowledge and worldview, can function positively or negatively within people's lives. It is important that the negative is researched in tandem with the positive. Some studies indicate a need to be wary of utilizing religion within a healthcare context (Paley, 2008). Such studies need to be placed alongside those highlighted in this chapter. However, medication has side effects, surgery has side effects, indeed any healthcare intervention or practice has a shadow side. The fact that religion and spirituality may function positively and negatively does not in any sense invalidate the need to research them or to practise them with care and theological and therapeutic discernment. Spirituality's engagement with science provides an opportunity critically to explore the ways in which these perspectives on the world can maximize our understanding and minimize any potential dangers. In conclusion, I propose that the most effective way to frame the relationship between religion and science in this context, or perhaps in any context, is through the concept of *hospitality*. In order to be hospitable to someone else you need to value, respect and honour the integrity of that person. If I'm trying to be hospitable to you, but at the same time I really want you to be like me, or even to be me, then we will never become friends. But if I approach you with confidence that what I have to bring to the table is important and will be valued, then we can commune. If you can come to me with the same confidence, then we can have a conversation, and in the end we may become friends. If we become friends then we can work together.

12

Do the miracles of Jesus contradict science?

MARK HARRIS

The problem of Jesus' miracles in the modern world

Jesus is universally acknowledged – even by those disinclined to religious belief – to have been a wise teacher who still has challenging and relevant things to say to us today. The stories of his miracle-working have fared less well, though, even among many practising Christians, some of whom find the miracles to be a more or less dispensable part of the Jesus tradition in comparison with the teachings. Much of this reticence towards miracle stems from the success of modern science, which is often held as having proved that miracles are (a) scientifically impossible and (b) a relic of a bygone primitive and gullible age. For these reasons, it's not unusual to hear the opinion voiced that science has disproved miracles in general and the miracles of Jesus in particular. And yet, as I hope to demonstrate in this chapter, such sweepingly sceptical generalizations are unsustainable in the light of the complexity of the relationship between miracle and science and in the light of the complexity of the miracle traditions of Jesus.

Hence, this chapter will look at one of the most divisive issues that arises for Christianity in the light of modern science, namely belief in miracles, and especially belief in the miracles of Jesus. The amazing ongoing successes of the natural sciences in explaining the world have meant that much of what previous generations of Christians believed easily has become more difficult for us, or so we often assume. I will examine this assumption by looking at the question that is so often asked when the subject of Jesus and miracles arises in discussion today: 'Do the miracles of Jesus contradict science?' As I hope will become clear, a considerable amount of ground needs to be cleared before we can address this question directly, and even then a simple 'yes' or 'no' will not suffice. The upshot is that on this particular topic of miracle, where the scope for conflict between science and religious faith might appear to be very high, we in fact find that the difficulties raised by science are relatively modest compared to other issues that arise from faith presuppositions and interpretation, issues that have been with us for very much longer than modern science.

I will spend some time looking at the problem of how to define miracle – including the ubiquitous modern understanding that a miracle is a transgression of the laws of nature – before suggesting (what I hope is) both a more sophisticated and a more pragmatic view, that a miracle defies easy definition because it's so reliant on an individual's presuppositions concerning science, faith and nature (all conveniently summed up by the blanket term 'worldview'). Nevertheless, I don't plan to move miracles out of reach, since I'll then move on to look at what we know of the miracles of Jesus, and at how rational–scientific attempts to explain them away largely miss the point. In other words, although the miracles of Jesus prove to be highly complex (rather like the definition of miracle itself), they remain intriguing and resistant to modern attempts at dismissing them.

In this way, this chapter will I hope go some way towards explaining the current state of play in research on the theology of miracle, and on the miracles of Jesus.

What is a miracle?

Before turning to Jesus, it's first necessary to ask, 'What is a miracle?' One of the old chestnuts of philosophy and theology, the definition of miracle is a long-standing problem with many solutions, only some of which place miracles in direct conflict with science. Hence, in spite of the modernist assumption that science has ruled miracles out of court, the truth is considerably more subtle.

The solution of David Hume (1711–76), the philosopher of the Scottish Enlightenment, is often taken as the starting point for defining a miracle, since his definition places miracles in direct conflict with science (or at least with 'the laws of nature'). For that reason, we will spend some time looking at Hume in detail. Hume's definition appears in Chapter X ('Of Miracles') of his *An Enquiry Concerning Human*

Understanding (1748) in an endnote which has become more famous than anything in the main text: 'A miracle may be accurately defined, *a transgression of a law of nature by a particular volition of the Deity, or by the interposition of some invisible agent*' (Hume, 2007, p. 127, Endnote [K] to 10.12). Here we see the ubiquitous modern understanding of miracle as a supernaturally caused event that violates a law of nature. And insofar as the laws of nature fall within the territory of the natural sciences, a miracle is in direct conflict with science. Hume's understanding of miracle is therefore a very sharp representation of the idea that science and religion are fundamentally opposed. There are problems, however. Hume's definition is heavily reliant on how the term 'law of nature' is to be understood. For one thing, 'law of nature' is an ancient metaphor that is so commonplace in our day (thanks to its adoption by modern science) that we've largely forgotten that it works by seeing the regularities of the natural world in terms of the statutes and ordinances that regulate human behaviour. But this does, of course, automatically put a negative spin on miracle. A miracle is the *abuse* of an otherwise regulated system – the triumph of disorder over order – and in keeping with the legal language, Hume describes a miracle as a 'transgression' or a 'violation' (Hume, 2007, p. 127, Endnote [K] to 10.12).

This negativity is a major weakness of Hume's definition, since it assumes a rigidity and a perspicuity to the laws of nature which are beyond the bounds of science and philosophy to prove. Hume's view of the natural world is one of strict predeterminism, a view that may have seemed obvious enough in the eighteenth and nineteenth centuries (when science was dominated by the clockwork determinism of Newton), but which hardly holds true today in the face of scientific developments which emphasize *indeterminism*, such as quantum mechanics, complexity, chaos and emergence. Moreover, despite a century or more of debate, philosophers of science are divided as to whether the laws of nature should be seen as *prescriptive* or *descriptive* of nature's regularities. Hume's definition of miracle rather assumes the former, that the laws are prescriptive, which means that it would be fatally wounded if the laws in fact turned out to be descriptive (i.e. provisional and approximate). Finally, in its reliance on the law of nature *metaphor*, Hume's definition of miracle is fundamentally inconsistent: the one who enforces the laws (i.e. God) is the same one who violates them.

In addition to these direct criticisms of Hume, indirect criticism can also be made on the grounds that an event need not contradict the laws of nature in order to be accredited as a miracle. C. S. Lewis provides one perspective on this point in his famous book *Miracles* (2012 [1947]), explaining that it is misleading to define a miracle as something that breaks the laws of nature, since the raw material of a miracle is always nature itself, which by definition will behave naturally. It's the *cause* of the event which is seen to be miraculous, he suggests, not so much its particular workings. So if, for example, God were to move an atom deliberately, then it might well precipitate an event that could be construed as 'miraculous', but once the atom has been moved it conforms perfectly to the laws of nature. As Lewis puts it:

> Nature digests or assimilates this event with perfect ease and harmonises it in a twinkling with all other events. It is one more bit of raw material for the laws to apply to, and they apply . . . If God creates a miraculous spermatozoon in the body of a virgin, it does not proceed to break any laws. The laws at once take it over. Nature is ready. Pregnancy follows, according to all the normal laws, and nine months later a child is born. (Lewis, 2012 [1947], p. 94)

We might question Lewis as to how precisely he imagines God creating a miraculous spermatozoon without breaking laws of nature (unless Lewis imagines a localized but otherwise fully *ex nihilo* act of creation at this point, which would seem to be stretching the definition of miracle somewhat). Indeed, Lewis's view does little to help us understand miracles which potentially involve far more than God moving an atom here or there (such as the resurrection of Jesus or the parting of the Red Sea). However, Lewis's emphasis on *God* as the cause of the miracle, over the question of whether or not it breaks a law of nature, is valuable. In other words, we should be careful not to get hung up on Hume's definition of miracle as 'a transgression of a law of nature', when the weight should really lie on 'by a particular volition of the Deity', the next part of Hume's definition.

Another example of why a miracle need not break a law of nature is provided by Colin Humphreys' study of the miracle texts of Exodus (*The Miracles of Exodus*, 2003). He provides naturalistic explanations for all of the miracle stories – including the plagues of Egypt and the spectacular parting of the Red Sea – interpreting them in terms of unusual but thoroughly natural phenomena that just happened to occur at the right time for the deliverance of the Israelites. The miracle is in the timing, as it were, but otherwise the stories are fully explicable scientifically and would not therefore be classified as miracles in Hume's definition. These are full-blown miracles that nevertheless obey the laws of nature, according to Humphreys.

Clearly, a considerable array of questions and qualifications are stacking up against Hume's definition of miracle as we reflect upon it, and we might thereby wonder why it enjoys such ubiquity today. The reason must be at least partly because Hume's definition of miracle serves the similarly ubiquitous notion that science and religion are totally opposed, confronting each other with mutually exclusive views of reality. For Hume's definition is highly confrontational: a miracle can only occur when one form of reality (the religious) violently invades the other (the natural); otherwise there are no points of contact between the two, and the natural is sufficient to explain all of the reality that we know.

I'll come back to the relationship between science and religion later, but for now (and in case it might be thought that I'm rejecting Hume's thinking on miracles), I'd like to underscore the *value* of what Hume says. As I mentioned earlier, the famous definition is actually found in an endnote to Hume's discussion of miracles, and in fact hardly captures the full force of what Hume is trying to say overall. For the main body of his text is far more concerned with establishing the kinds of evidence that one might need to accept as proof of a miracle, rather than with making a watertight definition for one. And indeed, mention of biblical texts such as Exodus just now, as sources of evidence for miracles, leads us on to a very important point made by Hume in his chapter, a point that is much less discussed than his definition of miracles but which is far more robust. As a thought experiment, Hume explores the kind of evidence we would need in order to believe in the resurrection of the famous Tudor queen, Elizabeth I, who the history books tell us died in 1603. Let's suppose that evidence comes to light that she died in 1600, and then came back to life, reigned for a further three years, and finally died a second time in 1603. What kind of historical testimony would convince us of her resurrection? Would the written evidence of doctors or of numerous courtiers of good character convince us? Certainly not. Hume's point is that no amount of witness evidence or testimony would convince us of the truth of this ridiculous story. Hume carefully avoids making the obvious connection with the resurrection of Jesus (and with the Gospel accounts as 'evidence'), but it's hard to avoid the conclusion that he is attempting to sow seeds of doubt there.

Notwithstanding Hume's scepticism towards evidence for a miracle, I believe that his approach actually makes possible some positive perspectives on miracle. My point is that no one today has anything invested in the resurrection of Elizabeth, so we are far more likely to assume the default position of scepticism in her case, that dead people simply do not come back to life. Christians are not, however, so sceptical when it comes to the resurrection of Jesus. But with Elizabeth no one stands to gain anything one way or another, so we are most likely to conclude there must have been some mistake or fraud. Hume's point is this: 'that no testimony is sufficient to establish a miracle, unless the testimony be of such a kind, that its falsehood would be more miraculous, than the fact, which it endeavours to establish.' In other words, for a potentially miraculous event to be accepted, the witness to it must be so unimpeachable that it is more plausible to believe in the reported miracle than to believe the witness might be wrong. In such a case, the falsehood of such a perfect witness would constitute a 'greater miracle' (as Hume puts it) than the reported miracle in question, and we would reject the greater miracle (the falsehood of the perfect witness) and accept the littler (the reported miracle in question). In practice, of course, few witnesses or testimonials are going to be so persuasive, especially if we can think of reasons why the witness might be biased by being predisposed towards belief in miracles. (Reports of religious miracles frequently fall into the 'biased' category, according to Hume, and should be treated with extra caution.) However, it's important to realize that Hume isn't ruling out miracles altogether, just pointing out that the odds against believing them are extraordinarily high, and stack up in relation to the difficulty of establishing reliable evidence of them. In a case such as the resurrection of Elizabeth, centuries in the past, and where we might think of any number of reasons as to why her courtiers and historians might want to glorify her by fabricating or amplifying such a story, then it will clearly be immensely difficult to establish strong and unbiased evidence, and we are likely to remain highly sceptical.

Hume is, I believe, absolutely right that we would always assume scepticism with an example such as the resurrection of Elizabeth. The case of Jesus is quite different, though, because while there are many sceptics, there are equally countless believers in his resurrection. We are left with the question of why so many are ready to believe in the resurrection of Jesus when the hypothetical parallel of Elizabeth would inspire widespread disbelief. The answer must rely on our individual (subjective) predispositions based on religious faith. In other words, belief in Jesus' resurrection, for so many Christians, does not rest on the witness testimony of the Gospels, which, by Hume's standards, would count as extremely weak and biased evidence. Rather, Christian belief in the resurrection of Jesus is crucially affected by individual worldview and predispositions towards faith. Many Christians profess to a personal and living relationship with Jesus which transcends and prefigures considerations regarding texts and evidence for a miracle. Hume didn't take such 'worldview' concerns into account; instead his account seems to assume that every assessment of a miracle tradition should be made indifferently and free of bias. We might wonder whether Hume himself was free from bias, especially when we take his notorious scepticism towards Christianity into account. Hence, we don't need postmodernism to realize that Hume's ideal 'objective' assessment is unrealistic (and probably impossible), which is the real weakness of Hume's account of miracle.

This leads us to emphasize the complex subjective factors at play in assessing and defining a miracle, despite Hume's attempt to arrive at a single objective definition. There is a balance to be achieved between subjectivity and objectivity. This balance will always involve a value judgement needing to be made concerning any given event that might be deemed miraculous, a value judgement that takes into account the supposed miracle's level of remarkability (which may or may not include a transgression of a law of nature), its theological significance and the nature of the evidence that supports it (including questions about how that evidence is to be interpreted). All of these factors could (and probably will) vary according to each individual's faith presuppositions; the factors are heavily subjective, in other words. But the balance must also allow for the possibility that miracle can be spoken of in meaningful ways that the whole Christian community can accept; the miracle must retain a degree of 'objectivity' to be a miracle, in other words.

The miracles of Jesus

There's no easier way to see how complex are the factors that come into play in assessing miracles than to look in detail at those surrounding the figure of Jesus. Here, our sources are predominantly the four Gospels. (It's worth noting, though, that the apocryphal gospels also include various miracle stories. Perhaps the best-known collection of miracle stories is that found in the *Infancy Gospel of Thomas*. Owing to the fact that this latter work serves to amplify a superhuman – and frankly rather scary – portrait of Jesus, it is not regarded as possessing any authenticity whatsoever.) Despite occasional claims that Jesus never existed in history, a theme much beloved of popular media, Jesus is in fact one of the best-attested historical figures of antiquity. He is also noted widely as a miracle worker, not only in the ancient Christian sources, but in some non-Christian sources too. Hence, whatever our predispositions for or against miracles, it's clear by any reasonable historical approach that Jesus was known as a miracle worker in his day, especially as a healer and exorcist.

However, Jesus' miracles are very diverse. If, as we saw above, the problem of defining miracle itself is complex, then the complexity increases when we look at how miracles arise in the story of Jesus. In fact, if we examine the story of Jesus as presented in the four canonical Gospels (Matthew, Mark, Luke and John), we see that miraculous happenings are reported at every stage of the story, from his birth to his death and resurrection, and at every stage in between (e.g. the temptations in the wilderness in Matthew 4 and the Transfiguration on the mountain in Mark 9.1–10). Multiple appearances of the risen Jesus are described at the ends of the Gospels, formidable miracles by any definition. Also, miracles proliferate during Jesus' public ministry before his crucifixion, both in quantity and in diversity. We see many stories of healings and exorcisms where human subjects are healed (and sometimes brought back from the dead, e.g. Luke 7.11–17), along with what are often referred to as 'nature miracles', where the remarkable element appears to involve divine manipulation of the natural or inanimate world (e.g. the feeding of the 5,000 in Mark 6.34–44 or the stilling of the storm in Mark 4.35–41). Some of these miracle stories are reported across several of the Gospels (notably the feeding of the 5,000, which is the only miracle in Jesus' pre-crucifixion ministry to feature in all four Gospels), while others are unique to just one Gospel (e.g. the raising of Lazarus in John 11). Gathering these features together, there's a great deal of complexity, with many disparate stories and traditions. Some stories are similar to those of other miracle workers known from this time in the Jewish and Graeco-Roman worlds (e.g. Hanina ben Dosa, who was, like Jesus, from Galilee), and some of the Jesus stories are similar to those told of legendary figures in the Old Testament (e.g. the ascension of Elijah in 2 Kings 2.11–12). Other Jesus stories are unique, and it seems that Jesus was distinctive in his time for having more stories told about his abilities as a healer and an exorcist than those of any other miracle worker.

The final point to note in gathering together all of the complexities involved in the miracle traditions surrounding Jesus is the way in which the traditions are passed on to us; that is, via the genre of *Gospel*. Although 'Gospel' is highly distinctive as a literary form – to the extent that scholars have wondered whether it might not in fact be altogether unique – it's also acknowledged that the four Gospels are related to a pattern of ancient biographies known as *bioi* ('lives'). And while there are similarities between this ancient genre and our modern biographical writing, there are also significant differences, especially if we come to the Gospels expecting to read a straightforward narrative of what happened in Jesus' life as it unfolded. For one thing, there is a great deal of divergence between the Evangelists over the contents and order of the story they tell of Jesus, including the interpretative glosses they place on it. There are even outright conflicts between them (e.g. the cleansing of the Jerusalem Temple takes place at the beginning of Jesus' ministry in John, but at the end in the Synoptic Gospels: Matthew, Mark, Luke). In fact, the Gospel writers are generally quite upfront about the fact that, whatever their claims to *historical* accuracy (e.g. Luke 1.1–4), their primary aim is to witness to the *theological* significance of Jesus (e.g. Luke 24.44–48). And we find that each Evangelist accounts for Jesus' theological significance in distinctively different ways, and each forges the story accordingly. In any case (and this point is not so obvious when reading the Gospels in modern English), the Evangelists were writing for people living 2,000 years ago from us, in very different cultures and contexts from our own. All of this means that we should take care before reading a story from the life of Jesus in the same vein as we might a factual 'objective' report from our modern day, say a newspaper report, since we would almost certainly miss a great deal of the significance of the story. This need for caution is nowhere more true than with the miracle stories contained in the Gospels, since the Evangelists use them in creative ways to interpret the significance of Jesus.

Considering the complexities surrounding miracles in general, and Jesus' miracles in particular, modern responses can be rather simplistic. Two opposite opinions are frequently encountered, both of which operate on the assumption that the texts give us straightforward historical reports of *what really happened*:

1 Rather than ask what the Evangelist is trying to tell us, it's assumed that the only relevant question to ask is, 'So this happened, and the people of the time believed it to be a miracle, but with our superior understanding of the world can it be explained in a non-miraculous way?'
2 The opposite response is to assume that faith in the miracle reported should be simple and trusting; nothing is to be achieved by questioning any detail of the story.

However, both attitudes, (1) and (2), miss the point of the story, and it's easy to show how. For instance, the Synoptics frequently tie Jesus' miracles to his central proclamation of the coming kingdom of God (e.g. Mark 1.39): the miracles demonstrate the veracity of this proclamation and help us to interpret it. It's no accident that many of the miracles described in the Synoptic Gospels are exorcisms: the point is that the kingship of the world is about to change from that of the devil to that of God. The miracles are therefore part of a sea change on a *cosmic* scale, with Jesus – as 'Son of God' (e.g. Mark 3.11) and 'Son of David' (Mark 10.46–51) – at its heart. In other words, the miracles are clues to a much bigger picture about the ultimate destiny of the universe. John's Gospel, on the other hand, shows very little awareness of the kingdom message, and instead connects its more select number of miracles (few of which parallel those in the Synoptics; there are no exorcisms in John, for instance) to the identity of Jesus directly: the miracles are now explicitly said to be 'signs' (e.g. John 2.11) to enable belief in Jesus himself as the source of 'life' (20.30–31). Faith is distinctive here, so while the Johannine miracles are designed to prompt faith in the observer or reader, the synoptic miracles often take place only *because of* the prior faith of the subject (e.g. Mark 10.52).

There are many more differences in emphasis between the four Gospels at every level, illustrating that they came together in different contexts and to support different aims. By and large Christians have seen the value of maintaining these distinctions over and against harmonizing them, because the Gospels are consistent with each other in maintaining an overall aim of spreading the message about Jesus and promoting faith in him. The many divergences do, however, create a tension which can be detected in the various miracle stories as much as in any other details of the stories. In other words, the miracle stories are not value-free, but invariably come with a deeper message tied to the motives of the Evangelist. If we simply assume that the miracle stories are straightforward historical reports we'll miss this.

Science, rationalization and the miracles of Jesus

We're finally in a position to turn to the question in the title of this piece, namely, do the miracles of Jesus contradict science? Hume's ubiquitous definition of miracle might indicate that the miracles of Jesus must *by definition* contradict science, but I have been arguing here that the situation is considerably more subtle, both in terms of how a miracle ought to be understood and in terms of the diversity of miracle traditions attributed to Jesus. In fact, the question of whether the miracles of Jesus contradict science can be answered accurately both in the negative and in the affirmative, as follows.

No, the miracles of Jesus do not contradict science. A good example is the miraculous catch of fish; this could be explained away as a coincidence (albeit a very unlikely one) whereby a large shoal of fish just happened to appear near the boat when Jesus told his disciples to cast their nets. This could even be said of the stilling of the storm, that it was a coincidence that the storm abated when Jesus commanded it to. Some of the rationalistic and naturalistic explanations of the miracles are closely related to this idea (e.g. that when Jesus was seen to walk on water, it was actually by means of a submerged sandbank), as is Colin Humphreys' suggestion mentioned above, that the miracles of Exodus are unusual natural phenomena that happened at just the right time. Related also are social-scientific studies of the miracles of Jesus, which point out the similarities between many of Jesus' deliverance and healing miracles and those of modern-day exorcists and folk healers (i.e. those who are recognized in their local society as having healing expertise while having no professional credentials). Research has established that taboos surrounding social exclusion and purity can result in psychosomatic effects in some cultures. Folk healers have been known to bring about amazing acts of healing by addressing the underlying social causes of such conditions. The suggestion is that Jesus was some kind of analogous folk healer in his own culture, thus providing an 'explanation' for many of his miracles in social-scientific terms. In short, there are several rationalistic avenues open to us if we wish to retain the miracle stories as historical events, without needing to believe that they must automatically contradict science.

Yes, the miracles of Jesus do contradict science. In spite of the rationalizations mentioned in the previous paragraph, there are some miracle stories that defy such an approach, most obviously those miracles where Jesus is said to raise dead people such as Lazarus back to life. There is no clear rationalistic approach available to us here, unless we simply deny that Lazarus was dead in the first place. And this raises a further difficulty for the rationalization approach of the previous paragraph, since it works by reading literally some of the details of the story but overlooking or denying others, especially those that relate to the deeper cause, God's special action. But it is not clear that a miracle story can be read in such a piecemeal fashion without effectively fabricating an alternative story. For instance, to explain Jesus' walking on the water as working by means of a submerged sandbank is to read one aspect of the story literally (how it appeared to the disciples) while denying another (the sense of what the story is trying to convey, namely that this man Jesus could do what only God could do). As a result such an approach completely misses the point of why the story is being told in the first place. Therefore, rationalization of a miracle story may force it into a scientific mould, but at some cost, since the deeper significance may thereby be lost. By definition, this deeper significance goes beyond the boundaries of science and reaches to the subjective faith component of miracle, encapsulated in its theological significance. In short, there are deeper levels of meaning at play beyond the simple question with which we are so often fixated in our modern scientific age, namely what actually happened in our (scientific) terms. I shall say more about this shortly.

So do the miracles of Jesus contradict science?

Yes and no. As I've suggested in the previous section, the question of whether the miracles of Jesus contradict science can be answered both in the affirmative and negative. This suggests that it's not a particularly penetrating question. Similarly, the question is unable to penetrate the issue we considered earlier about the definition of miracle, and the need to balance both subjective and objective factors. These factors come to the fore in the miracles of Jesus: the miracles are told in the Gospels primarily to attest to his significance, which is objective in the sense that it concerns the ultimate destiny of the world, but which can only be apprehended by faith; that is, subjectively.

The need to appreciate the objective–subjective dimensions of Jesus' miracles comes to the fore when we go beyond a surface reading and look at how they function in the Gospel narratives. Deeper currents become apparent, currents which have very little bearing upon science. For instance, the sea miracles parallel Jesus' exorcisms in illustrating his power over the forces of chaos. The sea miracles certainly indicate that Jesus has power over nature – he can do what only God can do in the Old Testament – but there is also a strong connection with Jesus' message (in the Synoptic Gospels) that the kingdom of God is at hand (or in other words, the ruler of the world is about to change from the devil to God). Therefore, the miracles have a cosmic significance in line with Jesus' teachings. A good example of the deeper significance of Jesus' miracles is provided by the story of the feeding of the 5,000. Here we find that the question of 'what really happened' can be answered in various rationalistic ways, for instance that it was an unexpected act of communal sharing by the crowd rather than a nature miracle where matter was literally multiplied. However, such rationalistic approaches make little impact on the deeper significance of the story, as revealed by the multiple layers of symbolism in the story regarding (for instance) the post-Easter mission of the Church, the Eucharist and the messianic banquet. To assume that the fundamental mystery behind the story can be answered by rationalizing it is to miss this. Therefore, in order to appreciate the multiple levels of meaning in the story for what they are (and to learn from them), it is necessary to see the search for a rationalistic–scientific explanation as just one way in which such a story is to be assessed. At least as important is the more subjective assessment that takes each story on its own terms and considers its impact on the assessor. Indeed, this subjective element is obvious in some of the miracle stories themselves, where the healing power comes not so much from Jesus as from the recipient's faith (e.g. 'Your faith has made you well', Matthew 9.22, ESV). The other side of the coin is provided by the people of Nazareth – those who have known Jesus for many years – whose lack of faith is such, we are told, that few miracles could occur there (Mark 6.1–6). Clearly then, the Gospels themselves bear out the objective–subjective duality of the miracle traditions. This means that the question at the title of this piece – 'Do the miracles of Jesus contradict science?' – must be honestly and accurately answered as 'yes' and 'no'.

How do science and religion relate, and what's it got to do with miracles?

Although much of the modern world appears to assume that science and religion are fundamentally opposed, I've tried to argue here that the truth is considerably more subtle, at least when we look at miracle traditions. If we were to try to characterize the relationship between science and religion based on what we've seen with miracles, we would have to admit that science has only a limited degree of purchase on religious claims for miracle, and that there's much more at stake than whether or not a miracle appears to break a law of nature. In any case, it's perfectly possible to accept a scientific explanation for any particular event while still maintaining that the event was a miracle in the sense of being specially caused by God.

Much comes down to the individual presuppositions we bring to the debating table, about our beliefs in the nature (and existence) of God and about how such a God might be said to work in the world. If we're determined to see an insurmountable barrier between the natural and supernatural (or to doubt the existence of the supernatural altogether), then we're likely to adopt an understanding of miracle like that of Hume, to prioritize the authority of science over religion and to believe that science and religion are independent at best or in violent conflict at worst. If, on the other hand, we are predisposed to be open-minded towards the possibility of divine action in the world, then we are likely to have a more nuanced view of the ability of science to pronounce on miracles. This means that we are likely to see science and religion as relating in more complex (or even integrative) ways, perhaps depending on a case-by-case analysis.

One issue that I haven't considered, but which is undeniably important in weighing up any potential miracle, is its 'cash value'. For Christians, the miracles of Jesus have a particularly high cash value compared to contemporary reports of miracles, simply because of the identity of the miracle worker and the role of the miracles in upholding that identity. With contemporary miracles, though, there could be a variety of responses. A miraculous salvation from a deadly situation is likely to be weighed up differently (and perhaps with greater seriousness) by a religious believer than a more trivial event, however remarkable the latter may seem. Other questions will also be relevant, such as why God might be said to accord miracles to some people and not to others who are equally deserving. Belief in miracles is therefore vulnerable to the questions that arise in discussing the problem of evil, namely why an omnipotent and good God should allow suffering in the world. Centuries of theological wrestling with these questions have found no easy answers. Hence, for myself, I must confess that my own belief in miracles, as a scientist and theologian, can change drastically depending on the type of miracle being reported, the nature and source of the evidence and what the supposed miracle might be said to achieve. I suspect that the wider relationship between science and religion is similar and like any supposed miracle needs to be taken on a case-by-case basis. This is not to give up on the questions, nor to affirm a totally relativistic or 'postmodern' view (as if 'anything goes'), but to insist that while such issues are resistant to easy categorization they should be debated all the same.

Conclusion

To conclude, I've given several reasons why the question in my title cannot be answered with a simple 'yes' or 'no'. I've tried to explain that this is not because we should be vague or non-committal about miracles, but because the question in the title does not penetrate to the heart of many miracle stories, especially those where the figure of Jesus is in view. Science has rather little to say on the matter compared to the dispositions of faith that we bring. I hope it's clear, then, that the question of whether the miracles of Jesus contradict science is actually rather different from the question, 'Do I believe in the miracles of Jesus?' This is a whole new question, for another day.

13

Can a scientist trust the New Testament?

N. T. WRIGHT

Epicureanism and scientism

Science covers many things. In modern Western science, two different narratives can become twisted together:

1 the continual exploration of the natural world, from distant stars to tiny particles;
2 the post-Enlightenment intellectual and social development of the modern Western world.

When these two stories get muddled, we have, not science, but scientism.

Scientism tries to extrapolate from the explosion of *scientific* knowledge proper to the belief that we now know far more about the moral, social and cultural world as well. This modern belief in *progress* results, not from observation of the natural world, but from a form of Epicureanism. For the Epicureans, the gods, if they exist, are far away and don't intervene. Instead, *the world works by itself*, evolving slowly and gradually. The great lie of today's scientism is that *science* has somehow proved Epicureanism.

What really happened was rather different. When Darwin went on his famous voyage and wrote his famous book, his findings did indeed demonstrate the high probability of the evolution of species. But Darwin's findings were seized upon by those who (for political and social reasons) already wanted to believe that the world simply developed itself, without divine intervention. Science became contextualized within, and then taken over by, scientism.

Scientism was then able to draw on the tradition of Descartes, approaching everything with systematic doubt. When asked if they could trust the New Testament, people in this tradition would say,

'Of course not. Where are the facts? Where is the scientific proof?' (They would have to say the same about Caesar's account of the Gallic War, or indeed Churchill's account of the Second World War.) Anyone combining the Cartesian tradition with Epicureanism would say, 'Not only can we not trust it; it says things which we know, a priori, to be false.' Anyone then combining these traditions with the doctrine of *progress* would say, 'And anyway, we know that these are only old fables, suitable for their time perhaps, but irrelevant to those who have escaped the dark night of superstition.'

But science, by itself, cannot adjudicate between different philosophical positions. Once all sides accept an implicit Epicureanism, then every advance in our understanding of natural causation looks like another nail in the coffin of the *god of the gaps*. This, very broadly, is where we are in Western culture – despite the newer movements in science itself, such as general relativity or quantum mechanics, which cast doubt precisely on some of the earlier *certainties*.

Different ways of knowing: how and why

There are, however, different kinds of *knowing*. Science studies the repeatable; history studies the unrepeatable. There are overlaps. Geology is, in a sense, the history of part of the natural world; so is astronomy. But we normally use *science* for disciplines that repeat experiments. With history, however, we depend on testimony, sometimes intentional (written accounts, etc.), sometimes accidental (archaeological remains, coins, and so on). Here the strict Cartesian, let alone the positivist, ought to worry: can we *know* what happened in the past, in the same way that we know the composition of a hydrogen molecule? Yet historians claim that they *do* know certain things: the fall of the Berlin Wall in 1989, or of Jerusalem in AD 70. And the crucifixion of a young Jew called Jesus by Roman soldiers outside Jerusalem in Passover week, probably in AD 33. And the fact that, shortly afterwards, his followers became convinced that he was alive again. As historians, we know all this as securely as we know about Jupiter's moons or the composition of the Cairngorm rocks.

Beyond science and history, there are different types of knowing. The most important things in life – music, faith, love, values, beauty, ethics, wisdom and hope – are not mere subjective fantasies. Somehow we have to hold together the work and findings of *science* with the things that really matter to us personally.

The former Chief Rabbi, Jonathan Sacks, offers a model for this. In *The Great Partnership* (2012), he proposes a formula: 'Science takes things apart to see how they work; religion puts things together to see what they mean.' History, too, enquires after meaning: not just 'what happened' but 'why'. Why did people start the First World War, drop the atom bomb, launch the Crusades, crucify Jesus? Without *meaning*, science and history alike become dry and bleak. By itself, *science* can tell you how to make a bomb, but not whether to drop it, or on whom. Scientism, however, is unwilling to allow for other spheres of enquiry, each with its own integrity. We must resist that intransigence. These spheres need to be held in partnership, in fruitful conversation.

All knowledge works by hypothesis and verification. An eternal dialogue takes place between our assumptions about how things make sense and our raw, unsorted encounters with the world. We form initial hypotheses, and test these against the data, reaching initial conclusions and modifying them in the light of subsequent experience or reflection. Sometimes, we have what Thomas Kuhn called a paradigm shift: new data doesn't fit, so the model itself tips over, generating a new paradigm. All this is well known; my point here is that all three areas I have mentioned, science, history and the worlds of religion, culture and art, proceed by this means. This is how we come to know *both* how things work *and* what they mean. That is the basis for our fruitful conversation.

Such a conversation can be found in the ancient Jewish tradition that Lord Sacks represents, and particularly (in my view) in the development of this which we find in Jesus and his first followers. And this brings us to the New Testament itself. I want to suggest not only that a scientist, *qua* scientist, can certainly trust the New Testament, but that the New Testament itself articulates modes of knowing which help us resist late-modern Epicureanism and the belief in automatic *progress*, and embrace wiser, more humane ways of thinking and being.

The roots of the New Testament are Jewish. The ancient Jewish vision of God is very different from that of Epicureanism. Israel's God, having made the world, continues in dynamic relationship with it, calling his people Israel for the sake of the world. This God declares *both* that he is the high and lofty one who inhabits a different space from us *and* that he dwells among his people. The Temple symbolized all this, being seen as the place where heaven and Earth met.

Israel's Scriptures express this belief from many angles; taken together, they form a narrative pointing ahead to a moment of truth. The New Testament claims that this moment arrived with Jesus of Nazareth, Israel's Messiah; that the God who made promises to Israel, promises involving creation itself, kept those promises in history, in Jesus. What might it mean to trust this testimony?

It means taking it seriously *as history*; which brings the problems into immediate focus. When people ask, 'Can a scientist – or any-one! – trust the New Testament?', they mean three things. First, can we trust the outline record of Jesus' public career? Second, can we believe in his *miracles*? Third, in particular, can we believe in his resurrection? Let me take these in reverse order.

Can we believe in the resurrection?

As to the resurrection: it isn't only modern science that *knows* that dead people don't rise. Here *scientism* overreaches itself, supposing it has discovered this for the first time. Believing that Jesus was raised from the dead always required a paradigm shift. Let us also be clear: the word 'resurrection' refers, not to *life after death*, but to a new, bodily life *after* whatever *life after death* there may be. The resurrection of Jesus is *not* about 'going to heaven', but about the launching of new creation within the ongoing old one.

So: can a scientist trust the New Testament's testimony about Jesus' resurrection? Scientism, of course, will say, 'Certainly not: we know things like that don't happen.' But a genuine scientist might say, 'Well, this is outside any other knowledge we have, so we will be suspicious; but when we look at all the evidence about the rise of the early Church, and the way it told these stories, different from anything in the imagination or mythology of the pagan world, different even from the pictures of "resurrection" within the ancient Jewish world, it seems as though we have to take the claim seriously.' And that would mean taking seriously the possibility that something quite new might have happened within the middle of human history, requiring other worldviews to be reworked around it. Either it is the new centre, or it is just a bizarre oddity – which almost certainly means it is nothing at all. But the question of whether you are prepared to treat it as the new centre is not a question that can be answered by science alone, or indeed history alone.

Can we believe in miracles?

When it comes to the other *miracles*, we note how that word itself has slipped over the years. Today people *hear* the word 'miracle' within an assumed Epicureanism: a normally absent divinity reaching in to the world, doing something bizarre, and then retreating again. The New Testament, however, asks us to consider two unexpected possibilities: first, that the God the Israelites invoked really was the creator of the whole world; second, that this God really had promised to come in person and bring the story of Israel, the story that would rescue the whole creation, to its unexpected and dramatic climax. The New Testament isn't suggesting that we fit these possibilities into our existing worldviews. It is offering these stories, knowing that they do not fit, but also knowing that, taken together, they constitute an invitation to reconstruct our worldviews in such a way that the story of Jesus – and everything else – will make a new kind of sense.

This, in principle, is the sort of thing scientists do regularly. Recently discovered prehistoric footprints in Norfolk will compel a new look at early human history. Science, *qua* science (as opposed to scientism), cannot pronounce on the unexpected. Genuine scientists welcome data that challenges existing hypotheses.

Can we trust the record of Jesus?

But I would not, myself, begin with the so-called 'miracles'. They are, in a sense, the icing on the cake. The place to start is either the resurrection itself, or the picture of Jesus' public career in the Gospels. Contemporary research on ancient Judaism continues, as I have argued elsewhere, to suggest that the Gospels are not the retrojection of later fantasy on to a falsely historicized screen. Their stories make sense in the world they claim to describe, offering a vivid portrayal of Jesus as a man of his time and yet a man exploding *into* that time with news: now, at last, Israel's God was becoming king *on Earth as in heaven*. Debate rages on the trustworthiness of this historical picture; but this points to the larger question, whether we can trust what Jesus himself said. Might it after all be the case that then and there, in the first century, in Palestine, the world's creator was starting to take charge of his world, challenging the principalities and powers, and doing so with the weapons not of revolution or military force but of forgiveness, healing and love? The real question faced by all of us, scientists included, is not just 'Could these things have happened?', but 'Could it be the case that there is after all a God who, having made the world, would come at last to sort it all out, and to do so in this way?' And that, of course, is not a question upon which the professional qualifications of a physicist, an astronomer, a botanist or any other scientist would entitle them to pronounce.

Nor can it be answered by simply saying 'yes' to all the particular questions about Jesus. Even if he did say and do those things, perhaps he was just a random freak. That brings us back to the resurrection. If Jesus had not been raised from the dead, his first followers would have concluded that it was all just a bizarre nonsense. The reason they didn't was that they believed, despite not having expected any such thing, that he had in fact been raised.

Jesus' resurrection, in fact, offers itself as a new centre not only of *what* we might know but of *how* we might know things. It doesn't fit into other philosophies, but then it doesn't claim to. It invites the question: what worldview or philosophy *would* you need to adopt if it *were* true? This is a regular scientific challenge: faced with data that doesn't fit the theory, you get a new theory. But with the resurrection something else seems to be going on, something which isn't just science and isn't just history, something in that larger, uncertain area about the meaning of all human life. What the resurrection offers is the introduction of a *new creation* – not a fresh creation out of nothing, but the rescue and revitalization of the old creation itself. It therefore offers a new mode of knowing, continuous with, and yet transcending, the modes appropriate for the present creation. The resurrection provides the bridge to speak the new word in language that can be heard in the old world, to invite the old world to recognize that new life has appeared even within its own sphere.

Conclusion

So: can the scientist – can any of us! – trust the New Testament? We certainly can't trust it to fit into our preconceived notions. But we can trust it to tell us about new creation, in such a way as to enable us to see that the old creation, with its own modes of knowing, is redeemed and taken up within it. And with that we can begin to rebuild *trust* in other areas of life as well – something we badly need at present.

But how? Here the New Testament puts one of its central proposals before us. It speaks of *power*: a power that works *through* the message about Jesus and his resurrection, and generates new modes of knowing and being. 'Trusting the New Testament' isn't a matter of a cool, detached appraisal. It means opening oneself to the source of life itself. The New Testament tells a story which invites, not spectators, but participants.

Trust is a larger category than scientific knowledge. It involves the natural, physical world, but also the world of history, and that larger, hard-to-define category which includes the things that really matter. So we come back where we began. The challenge of the New Testament is to discern the picture of God that we see in Jesus, and to learn to trust this God.

The real question, then, is not, 'Can *we* trust the New Testament, and the God of whom it speaks?' The question, really, is, 'Can this God trust *us* – to follow him, to reshape our worldviews around him, and make his glory known in the world?' Trust works both ways.

14

Questions for private thought
or group discussion

ERIC PRIEST

1 Introduction: Towards an integration of science and religion?

- What do you see as the relationship between science and religion, and why?
- If you are not a scientist, what do you imagine it is like to be one?
- If you are a scientist, describe your experiences of the nature of science.
- In what ways are the sciences and humanities part of an integrated whole?

2 God, science and the New Atheism

- Keith Ward describes God as a reservoir of unlimited energy. How would you begin to describe God?
- Give examples of scientific explanations for which experiments can or cannot be devised and which are or are not useful for improving the quality of life.
- Give examples of axiological explanations, identifying the four major elements mentioned by Keith Ward.
- What do you mean by God's consciousness and by God being eternal and unchanging?
- What is the relation between God and space–time?
- If you believe in God, what evidence can you give for that belief and in what sense is it rational?
- Describe ideas of New Atheism that you have come across; what arguments would you suggest against them?
- Describe examples from science of beauty, order and wonder.

- What are the values and limitations of science?
- What is your purpose as a human being?
- Describe examples of destruction and of creative emergence at work in the universe.

3 Natural law, reductionism and the Creator

- What is your reaction to the secularist scientific picture (SSP) described by Eleonore Stump at the beginning of her chapter?
- Are the laws of all the sciences reducible to those of physics and, if not, why not?
- Is everything determined by microphysical causal interactions?
- Compare the SSP and Thomist notions of natural law.
- Suggest examples, other than the example of autism, in which reductionism fails.
- Give reasons for rejecting reductionism.
- Discuss pros and cons of accepting the SSP or Thomist views for the ultimate foundation of reality.

4 The origin and end of the universe: A challenge for Christianity

- What is the role of God likely to be at the beginning and at the end of the universe?
- What do you mean by 'God sustains the universe'?
- In what ways can scientific work on the origin of the universe re-energize Christian theology?
- What is your reaction to the thought of a future universe that is almost empty and is steadily growing colder and darker for all eternity?
- In what ways can work on the long-term future of the universe give a renewed emphasis on new creation and Christian hope?

5 Universe of wonder, universe of life

- Describe examples from astronomy that give you a sense of wonder.
- If you were the first astronaut to stand on the Moon, how would you feel looking at the Earth in the sky?
- If life were discovered on other planets, would this change your view of humanity's significance?
- Would extraterrestrial life conflict or support theological ideas of creation?

- Do you think life forms beyond Earth would experience good and evil and would embrace a belief in God?
- Would Jesus redeem advanced life in other forms on other planets?
- If a multiverse were somehow detected, what effect would it have on your belief in God?
- Is there a purpose to the universe and, if so, what is it likely to be?

6 Evolution, faith and science

- How would you testify about the scientific standing of evolution against intelligent design?
- How does a person of faith view evolution?
- What are the elements of an understanding of evolution in harmony with Christianity?
- What do the first two chapters of Genesis and science tell us about the origin of the universe?
- What are the religious and scientific contributions to an understanding of the evolution of humanity?
- Would you regard yourself as a creationist and an evolutionist in the sense described by Dobzhansky?
- Do you believe in science because of your faith in God, as Consolmagno suggested?

7 Evolution and evil

- Is nature at war and characterized by evil?
- Is evil a central part of how nature works?
- Does evolution preclude the possibility of progress and of goodness?
- How can the ichneumon wasp be part of God's creation?
- How can a loving God allow the extinction of so many species in creation?
- Compare the roles of competition, natural selection and cooperation in evolution.
- Does evolution lack an aim or direction, and does it exclude goodness?
- For you, what conditions would a good explanation for suffering possess?
- Compare the explanations on pages 101–4 for the existence of evil, namely, the fall, freedom and evolution. Do either of them satisfy you?

- What do you think animal consciousness is like compared with our own?
- Compare the explanations on pages 104–9 for the existence of evil, namely, neo-Cartesianism, embodied intentionality and chaos-to-order. Do any of them satisfy you?
- Is pain the only or most effective way to keep us and animals from injuring ourselves?

8 Is there more to life than genes?

- Give several ways of describing the organization of a human.
- Describe examples of where opportunity determines behaviour and of where the environment could perhaps determine the expression of genes.
- Suggest examples of how genetic diversity has helped the survival or flourishing of our species.
- Suggest examples of how our brains function at an unconscious level and at a conscious level.
- Give examples of activity at the level of cells, at the level of organs and at the level of the whole person.
- Discuss examples of complex choices that need sound judgement, in which scientific insights and insights from the arts, philosophy or religion are important.
- Suggest examples of how religion has provided a basis for ethics and has helped us develop a purpose for living.
- Discuss examples of how humans operate at a rational level and at an intuitive level.
- How can we seek to live fulfilled lives where we flourish physically and emotionally?

9 Psychological science meets Christian faith

- Give examples of how an open-minded attitude has led to changes in your attitudes or ideas over the years.
- Discuss examples of a possible tension between psychological science and faith.
- Compare a theological and psychological description of mind and body.
- Compare biblical and psychological views on the relation between behaviour and belief.

- Discuss examples of where religion has in the past fostered horrors and goodness.
- Give examples of events in which one causes another and in which the two are only coincidentally correlated.
- What are the purposes and effects of prayer?
- What is your attitude to same-sex marriage?

10 Being a person: Towards an integration of neuroscientific and Christian perspectives

- Discuss the definitions of Peter Singer and John Harris for a person as someone with the capacity for enjoyment, for interacting with others, for having preferences or for valuing their own existence.
- Should only those with high-level cognitive functioning have moral rights and privileges?
- Should small babies be included in a definition of 'person'?
- Should those with advanced dementia be included in a definition of 'person'?
- Under what conditions would you sanction abortion?
- Compare the evidence for and against a dualistic philosophy of mind and body and holistic philosophy. Which appeals more to you?
- Is the sense of a conscious choosing self an illusion?
- Discuss Einstein's quote, 'The eternal mystery of the world is its comprehensibility.'
- Are persons a different kind of reality from matter?
- How do you react to the suggestion that you are made in the image of God?
- Discuss the statement, 'You love me, therefore I am.'
- What are the practical consequences of defining persons in the way suggested by John Wyatt?

11 From projection to connection: Conversations between science, spirituality and health

- In your experience, what has been the relationship between religion and healthcare?
- What are your own concepts of spirituality and religion and how do they differ?
- Describe your own experience of the benefits of religion for health.

- What are the main aspects of religion that may have a benefit for health?
- Compare a medical and a Christian definition of health.
- What can scientific enquiry tell us about religious experience and what can it not?
- Describe examples from your own experience of prejudice against religion in healthcare and examples where religion has functioned positively.
- Is it likely that human beings are hardwired for spiritual experience?
- In your experience, do children have an intrinsic sense of awe, wonder, connectedness and of the other?
- Compare the traditional Western view that individuals are free and autonomous with Hay's suggestion that they are by nature interconnected and relational.
- Discuss Hay's concept of *relational consciousness*. Does it appeal to you?
- Discuss consequences for modern secular society if indeed human beings are by nature relational.
- Suggest ways in which religion and healthcare can be more 'hospitable' to each other.

12 Do the miracles of Jesus contradict science?

- Discuss Hume's definition of miracle.
- How would you prefer to define a miracle?
- Read accounts of the following miracles in the Bible and discuss your reaction to them:
  - the crossing of the Red Sea (Exodus 13.17—14.29);
  - the healing of the blind man (Mark 8.22–26);
  - the healing of the paralytic (Mark 2.1–12);
  - the feeding of the 5,000 (Mark 6.30–44);
  - the miraculous catch of fish (John 21.1–14);
  - the stilling of the storm (Luke 8.22–25);
  - the raising of Lazarus (John 11.38–44);
  - walking on the water (Mark 6.45–52);
  - the resurrection (Mark 16.1–13).
- In each case, what is the deeper significance of the story beyond its simply being a miracle?
- Do you believe in the miracles of Jesus?

13 Can a scientist trust the New Testament?

- Has science proved that the world works by itself?
- Suggest some different philosophical positions for viewing the nature of reality; how would you adjudicate between them?
- Compare the different types of knowing that come from science, history, philosophy and religion. Are they completely separate or overlapping?
- Discuss the formula, 'Science takes things apart to see how they work; religion puts them together to see what they mean.' Suggest an alternative formula.
- Does all knowledge work by hypothesis and verification?
- Compare the Epicurean and Jewish visions of God.
- Can we believe in the resurrection?
- Can we believe in miracles?
- Can we trust the record of Jesus?

References and further reading

1 Introduction: Towards an integration of science and religion?

Bancewicz, R. (2015), *God in the Lab: How Science Enhances Faith*, Monarch, Oxford.

Barbour, I. G. (1997), *Religion and Science: Historical and Contemporary Issues*, HarperOne, New York.

Berry, R. J. (2001), *God and Evolution: Creation, Evolution and the Bible*, Regent College Publishing, Vancouver.

Dawkins, R. (1995), *River out of Eden: A Darwinian View of Life*, Weidenfeld & Nicolson, London.

Dostoyevsky, F. (1880), *The Brothers Karamazov*, The Russian Messenger, Moscow.

Draper, J. W. (1874), *History of the Conflict between Religion and Science*, D. Appleton and Co., New York.

Einstein, A. (1954), 'Science and Religion', *Ideas and Opinions*, Crown, New York, pp. 41–9.

Gould, S. J. (1989), *A Wonderful Life: The Burgess Shale and the Nature of History*, Norton, New York.

Gould, S. J. (1999), *Rocks of Ages: Science and Religion in the Fullness of Life*, Ballantine, New York.

Harris, M. (2013), *The Nature of Creation: Examining the Bible and Science*, Acumen/Routledge, Durham.

Harrison, P. (2007), *The Fall of Man and the Foundations of Science*, Cambridge University Press, Cambridge.

Harrison, P. (2015), *The Territories of Science and Religion*, University of Chicago Press, Chicago.

Hedley Brooke, J. (1991), *Science and Religion: Some Historical Perspectives*, Cambridge University Press, Cambridge.

Hull, D. L. (1991), 'The God of the Galápagos', *Nature* 352, pp. 485–6.

Huxley, T. H. (1863), 'On Our Knowledge of the Causes of the Phenomena of Organic Nature: Six Lectures to Working Men', *Collected Essays*, London, vol. 2, pp. 303–475.

McGrath, A. E. (2002), *The Dawkins Delusion*, SPCK, London.

McGrath, A. E. (2004), *The Twilight of Atheism: The Rise and Fall of Disbelief in the Modern World*, Rider, London.

McLeish, T. (2014), *Faith and Wisdom in Science*, Oxford University Press, Oxford.

Newton, I. (1999 [1687]), *The Principia: Mathematical Principles of Natural Philosophy*, trans. I. B. Cohen and A. Whitman, University of California Press, Berkeley and Los Angeles.

Numbers, R. L. (2009), *Galileo Goes to Jail and Other Myths about Science and Religion*, Harvard University Press, Cambridge, MA.

Oord, T. J. (2010), *The Polkinghorne Reader: Science, Faith and the Search for Meaning*, SPCK, London.

Paley, W. (1802), *A Natural Theology: Or, Evidences of the Existence and the Attributes of the Deity*, J. Faulder, London.

Peacocke, A. (1996), *God and Science: A Quest for Christian Credibility*, SCM Press, London.

Peirce, C. S. (1878), 'How to Make Our Ideas Clear', *Popular Science Monthly* 12, pp. 286–302.

Plantinga, R. J., Thompson, T. R. and Lundberg, M. D. (2010), *An Introduction to Christian Theology*, Cambridge University Press, Cambridge.

Polkinghorne, J. (1994), *Science and Christian Belief*, SPCK, London.

Robinson, J. A. T. (1963), *Honest to God*, SCM Press, London.

Sacks, J. (2012), *The Great Partnership: God, Science and the Search for Meaning*, Hodder & Stoughton, London.

Snow, C. P. (1959), *The Two Cultures*, Cambridge University Press, Cambridge.

Southgate, C. (2011), *The Groaning of Creation: God, Evolution and the Problem of Evil*, Westminster John Knox Press, Louisville, KY.

Templeton, J. M. and Herrman, R. L. (1998), *The God Who Would Be Known: Revelations of the Divine in Contemporary Science*, Templeton Press, West Conshohocken, PA.

Trigg, R. (2015), *Beyond Matter: Why Science Needs Metaphysics*, Templeton Press, West Conshohocken, PA.

Ward, K. (2008), *The Big Questions in Science and Religion*, Templeton Press, West Conshohocken, PA.

White, A. D. (1896), *History of the Warfare of Science with Theology in Christendom*, D. Appleton and Co., New York.

Wilkinson, D. (2010), *Christian Eschatology and the Physical Universe*, T&T Clark, London.

Wright, N. T. (2003), *The Resurrection of the Son of God*, SPCK, London.

Wyatt, J. (2009), *Matters of Life and Death: Human Dilemmas in the Light of the Christian Faith*, InterVarsity Press, Nottingham.

2 God, science and the New Atheism

Barbour, I. (1998), *Religions and Science*, SCM Press, Norwich.

McGrath, A. (2005), *Dawkins' God*, Blackwell, Oxford.

Peacocke, A. (1979), *Creation and the World of Science*, Oxford University Press, Oxford.

Polkinghorne, J. (1998), *Belief in God in an Age of Science*, Yale University Press, New Haven.

Ward, K. (1996), *God, Chance and Necessity*, Oneworld, London.

Ward, K. (2006), *Pascal's Fire*, Oneworld, London.

Ward, K. (2008), *The Big Questions in Science and Religion*, Templeton Press, West Conshohocken, PA.

Ward, K. (2014), *The Evidence for God*, Darton Longman & Todd, London.

Weinberg, S. (1977), *The First Three Minutes*, Andre Deutsch, London.

3 Natural law, reductionism and the Creator

Blackburn, S. (2002), 'An Unbeautiful Mind', *New Republic*, 5 and 12.

Boyd, R., Gasper, P. and Trout, J. D. (eds) (1993), *The Philosophy of Science*, MIT Press, Cambridge, MA.

Dupré, J. (1995), *The Disorder of Things: Metaphysical Foundations of the Disunity of Science*, Harvard University Press, Cambridge, MA.

Garfinkel, A. (1993), 'Reductionism', in Boyd et al. (eds), *The Philosophy of Science*, MIT Press, Cambridge, MA, pp. 443–59.

Hendry, R. F. (2010), 'Emergence vs. Reduction in Chemistry', in Cynthia MacDonald and Graham MacDonald (eds), *Emergence in Mind*, Oxford University Press, Oxford.

Hobson, P. (2004), *The Cradle of Thought: Exploring the Origins of Thinking*, Oxford University Press, Oxford.

Kitcher, P. (1993), '1953 and All That: A Tale of Two Sciences', in Boyd et al., *The Philosophy of Science*, MIT Press, Cambridge, MA, pp. 553–70.

Richards, F. M. (1991), 'The Protein Folding Problem', *Scientific American* 264, pp. 54–63.

Stump, E. (2003), *Aquinas*, Routledge, New York and London.

Stump, E. (2015), 'Natural Law, Metaphysics, and the Creator', in G. Buijs, G. Glas, A. Mosher and J. Ridder (eds), *The Future of Creation Order, vol. 1: Perspectives from Philosophy, the Sciences, and Theology*, Springer, New York.

Van Inwagen, P. (1995), *God, Knowledge, and Mystery: Essays in Philosophical Theology*, Cornell University Press, Ithaca and London.

4 The origin and end of the universe: A challenge for Christianity

Barbour, I. G. (1997), *Religion and Science: Historical and Contemporary Issues*, HarperOne, New York.

Barrow, J. D. (1993), *The Observer*, 7 May.

Davies, P. (2002), 'Eternity: Who Needs It?', in G. F. R. Ellis (ed.), *The Far Future Universe: Eschatology from a Cosmic Perspective*, Templeton Foundation Press, Radnor, PA.

Dyson, F. (1998), *Infinite in All Directions*, Harper & Row, New York.

Einstein, A. (1930), 'Religion and Science', *New York Times Magazine*, 9 November, pp. 1–4.

Einstein, A. (1998), Letter to Paul Ehrenfest, 17 January 1916, in *The Collected Papers of Albert Einstein: vol. 8: The Berlin Years: Correspondence, 1914–1918*, ed. R. Schulmann et al., Princeton University Press, Princeton.

Hawking, S. W. and Mlodinow, L. (2010), *The Grand Design*, Bantam, London.

Hermanns, W. (1983), *Einstein and the Poet*, Branden Press, Brookline, MA.

Hoffmann, B. (1972), *Albert Einstein: Creator and Rebel*, New American Library, New York.

Krauss, L. M. (2012), *A Universe from Nothing: Why There Is Something Rather Than Nothing*, Simon & Schuster, London.

Page, D. (1998), 'Hawking's Timely Story', *Nature* 333, pp. 742–3.

Palmer, J. (2011), 'Nobel Physics Prize Honours Accelerating Universe': <www.bbc.co.uk/news/science-environment-15165371>.

Perlmutter, S. (2003), 'Supernovae, Dark Energy, and the Accelerating Universe', *Physics Today* 56, no. 4, pp. 53–60.

Perlmutter, S. et al. (1999), 'Measurements of Omega and Lambda from 42 High-Redshift Supernovae', *Astrophysical Journal* 517, pp. 565–86.

Rees, M. (2003), *Our Final Hour*, Basic, New York.

Riess, A. G. et al. (1998), 'Observational Evidence from Supernovae for an Accelerating Universe and a Cosmological Constant', *Astronomical Journal* 116, pp. 1009–38.

Riess, A. G. et al. (2001), 'The Farthest Known Supernova: Support for an Accelerating Universe and a Glimpse of the Epoch of Deceleration', *Astrophysical Journal* 560, pp. 49–71.

Tipler, F. (1994), *Physics of Immortality*, Anchor, New York.

Wilkinson, D. (2010), *Christian Eschatology and the Physical Universe*, T&T Clark, London.

5 Universe of wonder, universe of life

Butt, R. (2009), 'The Vatican Ponders Extraterrestrials', *The Guardian*, 11 November.

Chappell, D. F. and Cook, E. D. (eds) (2006), *Not Just Science: Questions Where Christian Faith and Natural Science Intersect*, Zondervan, Grand Rapids.

Crowe, M. J. (2008), *The Extraterrestrial Life Debate: Antiquity to 1915*, University of Notre Dame Press, South Bend, IN.

Danielson, D. (2000), *The Book of the Cosmos: Imagining the Universe from Heraclitus to Hawking*, Perseus, Cambridge, MA.

Gingerich, O. (2014), *God's Planet*, Harvard University Press, Cambridge, MA.

Kaufman, M. (2009), 'The Religious Questions Raised by Aliens', *The Washington Post*, 22 December.

Miller, K. (ed.) (2003), *Perspectives on an Evolving Creation*, Eerdmans, Grand Rapids and Cambridge.

Planetquest (online compendium of exoplanet discoveries): <http://planetquest. jpl.nasa.gov>.

York, D. G., Gingerich, O. and Zhang, S.-N. (2012), *The Astronomy Revolution: 400 Years of Exploring the Cosmos*, CRC Press, Boca Raton, FL.

6 Evolution, faith and science

Augustine of Hippo (1982), *On the Literal Meaning of Genesis*, trans. J. H. Taylor, SJ, Newman Press, New York.

Boston, R. (1999), 'Missionary Man', *Church and State*, April.

Bunting, M. (2006), 'Why the Intelligent Design Lobby Thanks God for Richard Dawkins', *The Guardian*, 26 March.

Consolmagno, G. (2005), 'By Design', *Astrobiology*, September.

Darwin, C. (1859), *On the Origin of Species*, John Murray, London.

Dawkins, R. (1995), *River out of Eden*, Weidenfeld & Nicolson, London.

Dobzhansky, T. (1973), 'Nothing in Biology Makes Sense Except in the Light of Evolution', *The American Biology Teacher*, March.

Einstein, A. (1954), *Ideas and Opinions*, Crown, New York.

Giberson, K. (2008), *Saving Darwin: How to Be a Christian and Believe in Evolution*, HarperOne, New York.

Haught, J. (2001), *God After Darwin: A Theology of Evolution*, Westview Press, Boulder, CO.

Hull, D. L. (1991), 'The God of the Galápagos', *Nature* 352, pp. 485–6.

International Theological Commission of the Roman Catholic Church (2004), *Communion and Stewardship: Human Persons Created in the Image of God*.

Lebo, L. (2008), *The Devil in Dover*, The New Press, New York.

Miller, K. (1999), *Finding Darwin's God: A Scientist's Search for Common Ground between God and Evolution*, HarperCollins, New York.

7 Evolution and evil

Ayala, F. J. (2007), *Darwin's Gift to Science and Religion*, Joseph Henry Press, Washington, DC.

Barash, D. (1977), *Sociobiology and Behavior*, Elsevier North-Holland, New York.

Boyd, G. (2001), *Satan and the Problem of Evil: Constructing a Trinitarian Warfare Theodicy*, IVP Academic, Westmont, IL.

Darwin, C. R. (1856), Letter to J. D. Hooker, 13 July.

Darwin, C. R. (1858), 'Extract from an Unpublished Work on Species, Consisting of a Portion of a Chapter Entitled, "On the Variation of Organic

Beings in a State of Nature; on the Natural Means of Selection; on the Comparison of Domestic Races and True Species".'

Darwin, C. R. (1860), Letter to Asa Gray, 22 May.

Gould, S. J. (1988), 'On Replacing the Idea of Progress with an Operational Notion of Directionality', in M. Nitecki (ed.), *Evolutionary Progress*, University of Chicago Press, Chicago, pp. 319–38.

Hull, D. L. (1991), 'The God of the Galápagos', *Nature* 352, pp. 485–6.

Huxley, T. H. (1887), Draft of Manchester Address.

Huxley, T. H. (1888), 'The Struggle for Existence and Its Bearing on Man'.

Michod, R. E. (2000), *Darwinian Dynamics: Evolutionary Transitions in Fitness and Individuality*, Princeton University Press, Princeton.

Ohlson, K. (2012), 'The Cooperation Instinct', *Discover*, 29 December.

Polanyi, M. (1968), 'Life's Irreducible Structure', *Science* 160.

Polkinghorne, J. (2003), *Belief in God in an Age of Science*, rev. edn, Yale University Press, New Haven.

Tennyson, A. H. (1849), *In Memoriam A. H. H.*

Williams, G. C. (1988), 'Huxley's *Evolution and Ethics* in Sociobiological Perspective', *Zygon* 23, pp. 383–438.

8 Is there more to life than genes?

Gladwell, M. (2002), *The Tipping Point: How Little Things Can Make a Big Difference*, Back Bay Books, New York.

Longfellow, H. W. (2010), *The Song of Hiawatha*, Echo Library, Teddington.

Lucretius (Titus Lucretius Carus) (2007), *Book of the Nature of Things*, Penguin Classics, London.

Newman, J. H. (2014), *The Idea of a University*, Digireads.com, New York.

9 Psychological science meets Christian faith

Achtemeier, M. (2014), *The Bible's Yes to Same-Sex Marriage: An Evangelical's Change of Heart*, Westminster John Knox Press, Louisville, KY.

Brownson, J. V. (2013), *Bible, Gender, Sexuality: Reframing the Church's Debate on Same-Sex Relationships*, Eerdmans, Grand Rapids and Cambridge.

Cox, D. and Jones, R. P. (2014, 26 February), 'A Shifting Landscape: A Decade of Change in American Attitudes about Same-sex Marriage and LGBT Issues', Public Religion Research Institute: <www.prri.org/research/2014-lgbt-survey/>.

Dawkins, R. (1997), 'Is Science a Religion?', *The Humanist*, January/February, pp. 26–9.

Dawkins, R. (2001), 'Religion's Misguided Missiles', *The Guardian*, 15 September.

DeYoung, K. (2015), *What Does the Bible Really Teach about Homosexuality?* Crossway, Wheaton, IL.

Diener, E., Tay, L. and Myers, D. G. (2011), 'The Religion Paradox: If Religion Makes People Happy, Why Are So Many Dropping Out?' *Journal of Personality and Social Psychology* 101, pp. 1278–90.

Gagnon, R. A. J. (2001), *The Bible and Homosexual Practice*, Abingdon Press, Nashville.

Gould, S. J. (1997), 'Nonoverlapping Magisteria', *Natural History*, March, pp. 16–22.

Gushee, D. P., McLaren, B. D., Tickle, P. and Vines, M. (2014), *Changing Our Mind: A Call from America's Leading Evangelical Ethics Scholar for Full Acceptance of LGBT Christians in the Church*, David Crumm Media, Canton, MI.

Hummer, R. A., Rogers, R. G., Nam, C. B. and Ellison, C. G. (1999), 'Religious Involvement and U.S. Adult Mortality', *Demography* 36, pp. 273–85.

Johnson, W. S. (2006), *Time to Embrace: Same-Gender Relationships in Religion, Law, and Politics*, Eerdmans, Cambridge.

Jones, S. L. and Yarhouse, M. A. (2001), *Homosexuality: The Use of Scientific Research in the Church's Moral Debate*, InterVarsity Press, Downers Grove, IL.

Kahneman, D. (2011), *Thinking Fast and Slow*, Farrar, Straus & Giroux, New York.

Myers, D. G. (2008), *A Friendly Letter to Skeptics and Atheists: Musings on Why God Is Good and Faith Isn't Evil*, Jossey Bass Wiley, New York.

Myers, D. G. and Dawson Scanzoni, L. (2005), *What God Has Joined Together: The Christian Case for Gay Marriage*, Harper, San Francisco.

Myers, D. G. and Jeeves, M. A. (2002), *Psychology through the Eyes of Faith*, 2nd edn, Harper, San Francisco.

Pelham, B. and Crabtree, S. (2008), 'Worldwide, Highly Religious More Likely to Help Others: Pattern Holds throughout the World and across Major Religions', Gallup Poll: <www.gallup.com>.

Putnam, R. D. and Campbell, D. E. (2010, 17 October), 'Walking Away from Church', *Los Angeles Times*: <http://articles.Latimes.com/2010/oct/17/opinion/la-oe-1017-putnam-religion-20101017/>.

Rogers, J. B. (2006), *Jesus, the Bible, and Homosexuality: Explode the Myths, Heal the Church*, Westminster John Knox Press, Louisville, KY.

Wilson, K. (2014), *A Letter to My Congregation: An Evangelical Pastor's Path to Embracing People Who Are Gay, Lesbian and Transgender in the Company of Jesus*, Read the Spirit Books, Canton, MI.

10 Being a person: Towards an integration of neuroscientific and Christian perspectives

Crick, F. (1994), *The Astonishing Hypothesis: The Scientific Search for the Soul*, Simon & Schuster, London.

Harris, J. (1995), 'Euthanasia and the Value of Life', in J. Keown (ed.), *Euthanasia Examined*, Cambridge University Press, Cambridge.

Jeeves, M. and Brown, W. (2009), *Neuroscience, Psychology and Religion*, Templeton Press, West Conshohocken, PA.

Kuhse, H. and Singer, P. (1985), *Should the Baby Live?* Oxford University Press, Oxford.

Nagel, T. (2012), *Mind and Cosmos*, Oxford University Press, Oxford.

Pieper, J. (1997), *Faith, Hope, Love*, Ignatius Press, San Francisco.

Ridley, M. (2003), *Nature via Nurture: Genes, Experience and What Makes Us Human*, HarperCollins, London.

Tooley, M. (1983), *Abortion and Infanticide*, Oxford University Press, Oxford.

Wyatt, J. (2009), *Matters of Life and Death*, InterVarsity Press, Downers Grove, IL.

Zizioulas, J. (2004), *Being as Communion*, St Vladimir's Seminary Press, New York.

11 From projection to connection: Conversations between science, spirituality and health

American Psychologist: Journal of the American Psychologists Association (2003) (Special Issue: Health and Religious Beliefs) 58, no. 1.

Baumeister, R. F. and Sedikides, C. (eds) (2002), *Psychological Inquiry: An International Journal of Peer Commentary and Review* (Special Issue: Religion and Psychology) 13, no. 3.

CSHAD: Centre for Spirituality, Health and Disability, University of Aberdeen: <www.abdn.ac.uk/sdhp/centre-for-spirituality-health-and-disability-182.php>.

Hardy, A. C. (1965), *The Living Stream: A Restatement of Evolution Theory and Its Relation to the Spirit of Man. A Series of Gifford Lectures on Science, Natural History and Religion Delivered in the University of Aberdeen during 1963–1964*, Collins, London.

Hay, D. (2011), 'Altruism as an Aspect of Relational Consciousness and How Culture Inhibits It', in R. W. Sussman and C. R. Cloninger (eds), *Origins of Altruism and Cooperation*, Springer, New York, pp. 349–76.

Hay, D. and Nye, R. (2006), *The Spirit of the Child*, Jessica Kingsley, London.

Heelas, P. and Woodhead, L. (2005), *The Spiritual Revolution: Why Religion Is Giving Way to Spirituality*, Wiley-Blackwell, Hoboken, NJ.

Kirk, K. M., Eaves, L. J. and Martin, N. G. (1999), 'Self-transcendence as a Measure of Spirituality in a Sample of Older Australia Twins', *Twin Research* 2, no. 2, pp. 81–7.

Koenig, H. G., McCullough, M. E. and Larson, D. B. (2001), *Handbook of Religion and Health*, Oxford University Press, New York.

Levin, J. (2001), *God, Faith, and Health: Exploring the Spirituality–Healing Connection*, Wiley-Blackwell, New York.

McCullough, M. E. and Larson, D. B. (1999), 'Religion and Depression: A Review of the Literature', *Twin Research* 2, no. 2, pp. 126–36.

Miller, W. R. and Thoresen, C. E. (2003), 'Spirituality, Religion and Health: An Emerging Research Field', *American Psychologist* 5, no. 1, pp. 24–35.

Newberg, A. and D'Aquili, E. G. (2001), *Why God Won't Go Away: Brain Science and the Biology of Belief*, Ballantine, New York.

Paley, J. (2008), 'Spirituality and Nursing: A Reductionist Approach', *Nursing Philosophy* 9, no. 1, pp. 3–18.

Pargament, K. I. (2002), 'The Bitter and the Sweet: An Evaluation of the Costs and Benefits of Religiousness', *Psychological Inquiry* 13, pp. 168–81.

Schuman, J. and Meador, K. (2004), *Heal Thyself: Spirituality, Medicine and the Distortion of Christianity*, Oxford University Press, New York.

Sims, A. (1994), '"Psyche" – Spirit as Well as Mind?' *British Journal of Psychiatry* 165, pp. 441–6.

Sloan, R. (2006), *Blind Faith: The Unholy Alliance of Religion and Medicine*, St Martin's Press, New York.

Swinton, J. and Pattison, S. (2010), 'Moving Beyond Clarity: Towards a Thin, Vague, and Useful Understanding of Spirituality in Nursing Care', *Nursing Philosophy* 11, no. 4, pp. 226–37.

Thoresen, C. E. and Harris, A. H. S. (eds) (1999), 'Spirituality and Health', *Journal of Health Psychology* (Special Issue) 4.

Wilkinson, J. (1980), *Health and Healing*, The Handsell Press, Edinburgh.

12 Do the miracles of Jesus contradict science?

Corner, M. (2005), *Signs of God: Miracles and Their Interpretation*, Ashgate, Aldershot and Burlington, VT.

Eve, E. (2009), *The Healer from Nazareth: Jesus' Miracles in Historical Context*, SPCK, London.

Hume, David (2007), *An Enquiry Concerning Human Understanding*, ed. Peter Millican, Oxford World's Classics, Oxford University Press, Oxford and New York.

Humphreys, C. J. (2003), *The Miracles of Exodus: A Scientist's Discovery of the Extraordinary Natural Causes of the Biblical Stories*, Continuum, London and New York.

Keener, C. S. (2011), *Miracles: The Credibility of the New Testament Accounts*, Baker Academic, Grand Rapids, MI.

Lewis, C. S. (2012 [1947]), *Miracles: A Preliminary Study*, Collins, London.
Twelftree, G. H. (ed.) (2011), *The Cambridge Companion to Miracles*, Cambridge University Press, Cambridge.

13 Can a scientist trust the New Testament?

Sacks, J. (2012), *The Great Partnership: God, Science and the Search for Meaning*, Hodder & Stoughton, London.

Index

Did you know that SPCK is a registered charity?

As well as publishing great books by leading Christian authors, we also . . .

. . . make assemblies meaningful and fun for over a million children by running www.assemblies.org.uk, a popular website that provides free assembly scripts for teachers. For many children, school assembly is the only contact they have with Christian faith and culture, and the only time in their week for spiritual reflection.

. . . help prisoners to become confident readers with our easy-to-read stories. Poor literacy is a huge barrier to rehabilitation. Prisoners identify with the believable heroes of our gritty fiction. At the same time, questions at the end of each chapter help them to examine their choices from a moral perspective and to build their reading confidence.

. . . support student ministers overseas in their training through partnerships in the Global South.

Please support these great schemes: visit www.spck.org.uk/support-us to find out more.